Lecture Notes in Statistics 92
Edited by S. Fienberg, J. Gani, K. Krickeberg,
I. Olkin, and N. Wermuth

Mervi Eerola

Probabilistic Causality in Longitudinal Studies

Springer-Verlag
New York Berlin Heidelberg London Paris
Tokyo Hong Kong Barcelona Budapest

Mervi Eerola
Department of Statistics
University of Helsinki
PL 33 Aleksi 7
00014 Helsinki
Finland

Library of Congress Cataloging-in-Publication Data Available
Printed on acid-free paper.

© 1994 Springer-Verlag New York, Inc.
All rights reserved. This work may not be translated or copied in whole or in part without the written permission of the publisher (Springer-Verlag New York, Inc., 175 Fifth Avenue, New York, NY 10010, USA), except for brief excerpts in connection with reviews or scholarly analysis. Use in connection with any form of information storage and retrieval, electronic adaptation, computer software, or by similar or dissimilar methodology now known or hereafter developed is forbidden.
The use of general descriptive names, trade names, trademarks, etc., in this publication, even if the former are not especially identified, is not to be taken as a sign that such names, as understood by the Trade Marks and Merchandise Marks Act, may accordingly be used freely by anyone.

Camera ready copy provided by the author.
Printed and bound by Braun-Brumfield, Ann Arbor, MI.
Printed in the United States of America.

9 8 7 6 5 4 3 2 1

ISBN 0-387-94367-6 Springer-Verlag New York Berlin Heidelberg
ISBN 3-540-94367-6 Springer-Verlag Berlin Heidelberg New York

PREFACE

In many applied fields the explicit use of the notion 'causal dependence' seems to be avoided, perhaps because it is understood as a necessary relation, whereas reality in general does not correspond to constant conjunction of events. It is perhaps easier for statisticians than for many others, not familiar with probabilistic inference, to consider the question of causal dependence between events A and B, say, when B does not always follow from A. Also in the philosophy of science the notion of probabilistic causality is well-known. It means that the same circumstances produce variable outcomes with certain probabilities.

In this work we extend the classical theories of probabilistic causality to longitudinal settings and propose that causal analysis is essentially dynamic analysis. We also propose that interesting causal questions in longitudinal studies are related to causal effects which can *change in time*. We define the causal effect of an event as the effect of a change in the prevailing conditions due to the occurrence of that event at a particular time. The notion of 'conditions' or 'circumstances' in which events occur is therefore central in the longitudinal analysis of successive events forming a *causal chain*.

In some simple situations (e.g. when throwing a stone towards a window) we could in principle, by knowing the relevant causal laws (e.g. laws of physics) and the relevant circumstances (direction and speed of the stone, normal fragile glass), predict the response event (breaking of the window) in most details. The problems we encounter in statistical applications, in medicine, for example, are of course much more complicated and the causal laws are usually much less understood. Obviously we can then only in a very crude way model reality.

It is clear that already *explaining* the occurrence of an observed series of events requires knowledge not only about that particular situation but also general knowledge about relevant causal laws. In applications, the understanding of causal dependencies is most useful when trying to *predict* an outcome, be it the strength of a bridge or the survival of a patient after a radical operation. It is this predictive aspect which is our concern in this work. We therefore presuppose that the causal structure of the problem is reasonably well understood.

The theoretical basis of the proposed prediction method for longitudinal causal analysis is given in Chapter 2. A simple description of how statistics and probability theory are used in it is the following: Assume that a certain outcome can occur in a number of ways, or to use the language of series of events, there are many alternative "routes" (i.e., chains of events) which can lead to the same outcome (cf. Fig. 2.1 in Ch.2). A natural mathematical model for such a chain of events is a marked point process. The description of the dynamics in a sequence of events requires also the determination of those relevant conditions which produce the alternative events preceding the outcome. This can be done by statistical modelling. When these mechanisms have been determined, as well as we can in a particular problem, we can, by using the rules of probability theory, calculate the probabilities of the alternative routes, and eventually that of the outcome given these alternatives.

The explanatory aspect of causality corresponds then to the question: which of the possible routes was taken? (i.e., what happened?) Or, if this is not possible to answer: what was the most probable route (explanation)? The prediction aspect, on the other hand, corresponds to the question: which routes are still possible, and given these alternatives, what is the probability of the outcome?

The proposed prediction method provides a very general framework to study the dynamics and the magnitudes of causal effects in a series of dependent events. The question of relevant conditions and causal laws related to a particular causal problem is not primarily a statistical problem but merely a question of subject matter. However, our ability to estimate genuine causal effects from data is of course crucially dependent on our knowledge about these relevant conditions.

Finally, when introducing a statistical method it is important to test its usefulness with real data sets. Two examples are considered in Chapter 4. Similarly, it is important to test the accuracy of results in a particular application. For this purpose, the asymptotic variances of the prediction probabilities are derived in Chapter 3. The sensitivity of the method on model choice is considered in the end of Chapter 4, as well as outlines for its further uses.

Acknowledgements. I am grateful to Elja Arjas who initially suggested using the notion of a prediction process in causal analysis to me. He is also my co-author in a previous paper concerning this topic. I thank Niels Keiding for letting me use the transplantation data as an example and for several discussions. Pekka Kangas skillfully planned the initial computational framework. Juha Puranen and Martti Nikunen helped me with practical advice concerning the graphics and the output. The Statistics Departments at the Universities of Helsinki and Oulu provided good facilities for work. The financial support of the Academy of Finland is gratefully acknowledged.

May, 1994

Mervi Eerola

CONTENTS

PREFACE

1. FOUNDATIONS OF PROBABILISTIC CAUSALITY 1
   1.1 Introduction 1
   1.2 Historical aspects of causality 2
   1.3 Probabilistic causality 4
      1.3.1 I.J. Good: A quantitative theory of probabilistic causality . . . 5
      1.3.2 P. Suppes: A qualitative theory of probabilistic causality 6
   1.4 Different interpretations of probability in causality 7
      1.4.1 Physical probabilities 8
      1.4.2 Epistemic probabilities 9
   1.5 Counterfactuals in causality 12
   1.6 Causality in statistical analysis 13
      1.6.1 Randomized experiments 13
      1.6.2 Independence models 19
      1.6.3 Dynamic models 24
   1.7 Discussion 26
2. PREDICTIVE CAUSAL INFERENCE IN A SERIES OF EVENTS . . 29
   2.1 Introduction 29
   2.2 The mathematical framework: marked point processes 31
   2.3 The prediction process associated with a marked point process . . . 33
   2.4 A hypothetical example of cumulating causes 37
   2.5 Causal transmission in terms of the prediction process 42
3. CONFIDENCE STATEMENTS ABOUT THE PREDICTION PROCESS 44
   3.1 Introduction 44
   3.2 Prediction probabilities in the logistic regression model 45
   3.3 Confidence limits for μ_t using the delta-method 47
   3.4 Confidence limits for μ_t based on the monotonicity of hazards . 51
      3.4.1 Confidence limits for the hazard in the logistic model 52
      3.4.2 Stochastic order of failure time vectors 53
   3.5 Discussion 56
4. APPLICATIONS 57
   4.1 Multistate models in follow-up studies 57
   4.2 Modelling dependence between causal events 58
   4.3 Two applications 60

        4.3.1 The Nordic bone marrow transplantation data: the effects of CMV infection and chronic GvHD on leukemia relapse and death . . . 60
        4.3.2 The 1955 Helsinki cohort: the effect of childhood separation on subsequent mental hospitalisations 80
   4.4 Sensitivity of the innovation gains on hazard specification 94
   4.5 Discussion 101
   4.6 Computations 105
   4.7 Further uses of the method 105
        4.7.1 Quantitative causal factors 105
        4.7.2 Informative censoring and drop-out 107
5. CONCLUDING REMARKS 109

BIBLIOGRAPHY

APPENDICES 1-2

1. FOUNDATIONS OF PROBABILISTIC CAUSALITY

1.1 Introduction

The notion of causality is in many respects fundamental in scientific reasoning. Let us only mention some connotations that a causal relation between two events, cause and effect, may have:

(i) A cause precedes the effect;

(ii) Causes and effects are evidential to each other;

(iii) Causes explain effects;

(iv) A cause is a means to an end: the effect;

(v) A cause has predictive power concerning the effect.

The items (ii)-(v) implicitly include the assumption that there is a temporal relation between a cause C and its effect E which is stated in (i). Sometimes in human studies the means-end aspect of causality has a teleological or finalistic interpretation: goal-setting predetermines the means to reach the goal, and in this sense a cause and its effect have a converse relation than in the natural sciences, for instance. Human beings are capable of anticipating their future and of adapting their present behaviour to it (e.g. Eskola, 1989). In *event-causality*, which is our topic here, the temporal precedence of a cause seems, however, to be generally accepted, and it will be our starting point.

Unlike statistical dependence or association, causal dependence is *directed*; i.e., C causing E, or the occurrence of E depending on the occurrence of C. In some cases an event can have both the role of a cause and that of an effect: being first an effect, E may later act as a cause to another effect. This is the typical structure of a *causal chain*.

In (i)-(v) above time is present in different ways. In (ii), (iv) and (v) the effect is not assumed to have occurred yet but there are reasons to anticipate its occurrence since the cause has already taken place. In contrast, in (iii) also the effect is assumed to have occurred already, and we are looking for an explanation for its occurrence, that is, *why* did E occur? In this sense, causality has both *predictive* and *explanatory* aspects. These two aspects of causality are interrelated, but they correspond to different ways of conditioning on the observed history. The explanatory aspect is related to what is called *token* causality ("What was (were) the cause(s) of E?") in the case of *singular* events in the philosophical literature. The predictive aspect, on the other hand, corresponds to *generic* events and statements of the type "C causes E" (e.g. smoking causes lung cancer), meaning that "normally" or "frequently" events of type C are followed by events of type E. In this work we shall restrict ourselves to the predictive aspect of causality. Instead of a simple relation "C causes E", we consider a series of events, a causal chain. Despite the admittedly complicated philosophical problem of causal dependence, the aim of this work is to present a very concrete and applicable method for the statistical analysis of causal effects in an observed series of events.

Longitudinal studies or event-history analysis offers in many ways an ideal design for causal interpretations because in full event-histories the exact ordering of events is known. We may then ask more detailed questions, like "*When* did the cause happen? At what age?", for instance, or "Does a late occurrence of a cause change the prediction of the effect compared to an early one?" If there is a component cause acting in conjunction with other component causes, which together produce the effect, we have even more complex temporal relations between the events. It is well-known in the literature of causal modelling that different underlying causal structures can yield to exactly the same statistical models. It is impossible to distinguish between them if the ordering of the causal events is not known.

In this work we consider a dynamic approach to causal analysis in longitudinal studies along the lines of Arjas and Eerola (1993). The sequence of causal events is formulated as a stochastic process, more specifically as a marked point process, and the influence of a cause is expressed by the difference in the prediction probabilities of the effect given that the cause has just occurred, compared to the hypothetical situation that it did not occur. In Chapter 1 we shall first review the literature of probabilistic causality and in Chapter 2 we present the concept of a prediction process and its connection to causality. Statistical properties of the prediction probabilities are considered in Chapter 3, and in Chapter 4 we illustrate the ideas with some real data examples.

1.2 Historical aspects of causality

Causal relation between events played a central role in the philosophy of scientific reasoning of two important philosophers in the eighteenth century: David Hume and Immanuel Kant. The deterministic view of the time, which was supported by the success of the natural sciences, is formulated in Kant's *universal law of causation*. It implies that every event has a deterministic cause. In his criticism against inductive reasoning Hume (1748, 1771) argues that even though causal dependence plays a fundamental role in anticipating future events in daily life, there can be no a priori knowledge of such a dependence. Causal relationship between distinct events is not a logical nor a necessary relationship, but a factual relationship which must be judged by experience. Hume (1771) writes: "There is no necessity in causal relation. Necessity is something that exists in our mind, not in objects." Single events do not contain information about general causal dependence. It is the conjoined occurrence of the same type of events that raises the impression of causal dependence in our minds. Causality, according to Hume, is thus a mental image in human mind. There are, however, no grounds to expect that a particular causal relation should persist in time; i.e., that the world would remain the same in future. In Hume's view, this is the fundamental dilemma of inductive reasoning. In Kant's (1781, 1789) representation of all synthetical principles the proposition that nothing happens by chance is an a priori law of nature, and causality can be justified a priori even if singular causal relations are based on experience. Causality is one of Kant's basic categories and, in his view, an indispensable part of human understanding.

Since Hume's regularity theory of causality has been so influential, we shall have a more detailed look at the postulates in Treatise of Human Nature (1748). He wrote "...we may define a cause to be an object followed by another, and where all the objects, similar to the first, are followed by objects similar to the second. Or, in other words, if the first object had not been, the second never had existed". Hume's definition consists of three essential characteristics: *constant conjunction* (events of type C are always followed by events of type E; i.e., C is a sufficient cause to E), *locality principle* (causal events must be contiguous in time and space), and *succession in time* (a cause must precede the effect). The second part of Hume's definition, which shows that C is also a necessary cause of E, says that if the former event (C) had not occurred, neither would the latter (E) have existed, is a *counterfactual* statement. We shall consider the role of counterfactuals in causality in section 1.5 below.

Later developments of causality have shown several weaknesses in Hume's theory. First, it offers no means to distinguish between accidental generalizations and law like causal relations. If two events are observed to appear together, and they are contiguous, then by Hume's definition, they are causally related. Obviously, this is not a satisfactory principle. It is easy to find examples in which two completely unrelated phenomena appear together because of a common factor, for example. Ducasse (1951) therefore argues that Hume's definition merely raises the question whether two events are causally related, but it cannot be a definition of causality. Another too simplified assumption is that an effect can have only one cause. If there are several alternative causes which produce the same effect, then the first and second parts of Hume's definition cannot be true at the same time. Furthermore, causes and effects are constantly conjoining only *in certain circumstances*; i.e., C is a sufficient cause to E under certain additional conditions $A_1, A_2, ..., A_k$ which are called background conditions. Mackie (1973) has introduced the concept *insufficient but non-redundant part of an unnecessary but sufficient cause* (INUS-condition) for the case in which there exist several components in a sufficient cause which together are needed to produce the effect. A cause is thus always determined in a specific *causal field*.

Finally, we mention briefly John Stuart Mill's (1843) three basic rules for inventing and justifying causal relations among events in *experiments* because, as we shall see later on, the principles of randomized experiments play a fundamental role in the statistical approaches of analysing causal effects of several treatments. Mill's first rule is the *method of agreement* which says that if two or more instances of the event E have only one circumstance in common (C) then C is a sufficient or necessary cause for E: The *method of concomitant variation* concerns the variation of two phenomena: if E varies always when C varies, then C is E's sufficient or necessary cause. Finally, the *method of difference* resembles most the inference in statistical experiments: If the event E and its complement $\bar{E}$ differ only with respect to the factor C then C is E's sufficient or necessary cause, or an indispensable part of the cause.

1.3 Probabilistic causality

Causality has in everyday talk usually a weaker meaning than deterministic (strict) causality. We may say that ice on the road causes car accidents without meaning that in every single case this would happen. Similarly, we believe that smoking causes lung cancer, but not every smoker is assumed to develop lung cancer. Yet this is one of the most widely accepted examples of causal dependence in medicine. The interpretation of such statements is merely that the occurrence of the effect is highly, or at least more *probable* when the cause has already occurred than without it. In contrast to deterministic causality, in which the same cause is always followed by the same effect, in indeterministic causality the same cause may variably bring about different effects under the same conditions (e.g. female-male inheritance in genic transmission). Causal processes are then assumed to be of truly non-deterministic kind. Indeterminism in causality violates the invariability assumption which is a central part of causal laws (e.g. Nagel, 1961) and it is therefore not accepted by many philosophers. On the other hand, for example, Suppes (1984) considers indeterministic (or probabilistic) causality as the only form of causality which corresponds to our everyday interpretation of causal dependence. Mellor (1988) also points out that the connotations of causality (cf. Section 1) do not necessitate its strict deterministic definition.

Laplace, in his memoir from 1774, was probably the first who presented a definite methodology for estimating the probabilities of the causes that have produced an observed event. He used probabilistic considerations to infer whether the observations were results of a constant cause or simply due to random fluctuations or errors in measurement. As a simple example he showed, by detailed calculations, that the probability that the sum of heights of a barometer measuring the pressure of the atmosphere at nine o'clock in the morning exceeds by 400 millimeters the sum of the heights measured at four in the afternoon, is extremely close to one. He then inferred that there must be a constant cause behind the diurnal variation, and it cannot be a consequence of random fluctuations or errors in measurement. Laplace was, however, convinced that the probabilities in causality arise from our ignorance of true causal mechanisms, not from an inherent indeterminism of nature.

In probabilistic terms the strict assumption of constant conjunction by Hume is expressed as $P(E|C) = 1$ ("a cause C is always followed by the effect E"). Correspondingly, the counterfactual statement $\bar{C} \to \bar{E}$ is expressed as $P(\bar{E}|\bar{C}) = 1$, or equivalently, $P(E|\bar{C}) = 0$. Deterministic causality can then be viewed as a limiting case to the *frequent* conjunction requirement in which we let $0 < P(E|C) \leq 1$. Obviously the extreme case $P(E|C) = 0$ does not fulfil the requirement of frequent conjunction.

What would then be a "high enough" probability of E given C in order that C could be regarded as a cause for E? In theories of probabilistic causality it is assumed that the occurrence of a cause *raises* the probability of the effect. This suggests that there is a reference probability to which the *new* probability of E (after C's occurrence) is compared. This means that the occurrence of C raises the probability of E from the level it *has been* or *would have stayed* without C's occurrence. This is undoubtedly a dynamic question and the other part, "would have stayed", is a counterfactual if C actually occurred. In order to say that C raises the probability of E we should know what the probability of E is before C's occurrence.

Probabilistic theories offer important improvements to Hume's theory. In what follows we shall present in more detail two well-known philosophical theories of probabilistic causality: by Good (1961, 1962) and by Suppes (1970). Their approach is in many ways different. For example, Good's theory is aimed at quantitative causal reasoning whereas Suppes (1988) is doubtful about the usefulness of quantitative causal reasoning in general. They also consider different reference probabilities. In neither theory is the dynamic aspect, however, explicitly taken into account. Different interpretations of the probability concept are introduced in Section 1.4. In Section 1.5 the idea of counterfactuals in causality is presented and in Section 1.6 principles of causality and statistical analysis in certain applied fields are reviewed.

1.3.1 I.J. Good: A quantitative theory of probabilistic causality

I.J. Good (1961/1962) discriminates among the *tendency* of one event C to produce a later event E (denoted by $Q(E:C)$) and the *extent* to which C *actually caused* E (denoted by $\chi(E:C)$). The former relates to the question of predicting the occurrence of the effect when the potential cause C is present, whereas in the latter case also the effect E has already occurred, and in this single case, a quantitative measure for the extent of causal dependence is looked for. Good considers the latter case to be more difficult to define, whereas a measure for C's tendency to produce E can be given in terms of the probabilities $P(E|C), P(E|\bar{C})$ and $P(C)$. In each of these probabilities the state of the universe U (i.e., the prevailing conditions) and all true laws of nature are assumed to be included as a condition in addition to C. Good emphasizes that $Q(E:C)$ refers to the *tendency* of some event C to cause another event E, not the probability that C is *the* cause of E.

Good proposes that C's tendency to produce E be measured by

$$Q(E:C) = ln(\frac{1 - P(E|\bar{C})}{1 - P(E|C)})$$

i.e., as the logarithmic difference of the complements of the basic conditional probabilities $P(E|\bar{C})$ and $P(E|C)$. This expression does not depend on $P(C)$. This is important in Good's opinion since in an experiment a randomizing device is usually used to determine the probability of C. If the experiment is repeated many times to find out the extent to which C causes E, the probability of C should not affect these experiments. This also assures that there is no common cause producing both C and E.

As a quantitative measure of the strength of causal dependence $Q(E:C)$ has values in $[-\infty, \infty]$ and it increases when $P(E|C)$ increases, which seems a natural requirement. Since $P(E)$ is a weighted average of the probabilities $P(E|C)$ and $P(E|\bar{C})$, the role of the probability $P(C)$ can be obtained from

$$P(E) = P(C)P(E|C) + P(\bar{C})P(E|\bar{C}).$$

It is weighting the conditional probabilities by the probability of the cause in the population. In a controlled experiment, which Good considered, the spread of the cause is in the hands of the experimenter and thus not of interest as a random element. Finally, $Q(E:C)$ can be interpreted as the weight of evidence against C provided the nonoccurrence of E, which in Good's notation is $W(\bar{C}:\bar{E})$. It can be taken as a Bayesian likelihood ratio.

1.3.2 P. Suppes: A qualitative theory of probabilistic causality

Suppes' (1970) qualitative theory of probabilistic causality is perhaps the best known in the literature of probabilistic causality. Suppes describes his theory in the following way: "...Roughly speaking, the modification of Hume's analysis I propose is to say that one event is the cause of another if the appearance of the first event is followed with a high probability by the appearance of the second, and there is no third event that we can use to factor out the probability relationship between the first and second events". Suppes' main objection to classical deterministic causality was the assumption of constant conjunction. This requirement, he claims, is not in concordance with our everyday talk and experience of causality.

In the original definition Suppes considers the causal events as realizations of stochastic processes, and their occurrence times are included in the formal characterization of the theory. Then $P(E_t|C_{t'})$ is the conditional probability of event E occurring at t, given the occurrence of C at t', $t > t'$. The time aspect is, however, present only in the form $t' < t$. In later versions (e.g. Suppes, 1984) the explicit reference to time has been left out.

In Suppes' definition the event $C_{t'}$ is a *prima facie* cause of the event E_t if and only if

(i) $t' < t$,
(ii) $P(C_{t'}) > 0$,
(iii) $P(E_t|C_{t'}) > P(E_t)$.

To distinguish between accidental generalizations and true causal dependencies a prima facie cause is called *spurious* if there is a partition $\{B_{t''}\}$ into events $B^i_{t''}$, $t'' < t' < t$ such that

(i) $P(C_{t'} B^i_{t''}) > 0$,
(ii) $P(E_t|C_{t'} B^i_{t''}) = P(E_t|B^i_{t''})$, for some i,
(iii) $P(E_t|C_{t'} B^i_{t''}) \geq P(E_t|C_{t'})$, for some i;

otherwise C is a *genuine* cause of E. These conditions rule out the case in which an event, temporally preceding both the postulated cause and the effect, may have caused both C and E.

A prima facie cause is said to be *direct* if there is no partition $\{A_{t'}\}$ into events $A^i_{t'}$ temporally between $C_{t''}$ and E_t such that

(i) $t'' < t' < t$,
(ii) $P(C_{t''} A^i_{t'}) > 0$,
(iii) $P(E_t|C_{t''} A^i_{t'}) = P(E_t|A^i_{t'})$, for some i;

otherwise it is called an *indirect* cause of E_t. This corresponds to the case in which an intervening event A^i carries over the effect of C. As we shall see, the principles of ruling out spurious causes and finding direct and indirect causes have been used extensively in causal modelling in the social sciences.

1.4 Different interpretations of probability in causality

Among philosophers involved with statistical inference there has been much discussion about what different interpretations of the probability concept mean in the context of probabilistic causal statements. The notion of probability is usually classified as *physical* or *epistemic*, and this distinction is also present in the philosophical literature of causality. The basic difference between physical and epistemic probability concepts is in the role of human knowledge or information. Physical probabilities are assumed to exist irrespective of human consciousness whereas epistemic probabilities are always defined with respect to the state of knowledge of the interpreter. In causality, the difference of interpretations is connected to the question of how "final" the causal statements are considered. In the following we compare the notions of physical and epistemic probabilities in the light of causality. The presentation is by no means comprehensive because it serves only for the purposes of this study. For a more comprehensive presentation, we refer, for example, to Kyburg (1970).

1.4.1 Physical probabilities

Probabilistic propositions of physical probabilities concern objective facts in the physical world which are determined by natural laws, and which do not change as a consequence of improved understanding. In causality this means that causal relations exist in nature independently of our knowledge about them. In the frequentist approach probability is measured by relative frequencies which are empirical outcomes of repeating experiments or phenomena. The probability of an event is measured by the frequency of its occurrence in a series of experiments (Venn, 1886, von Mises, 1928, Reichenbach, 1949). In the long run these frequencies tend to stabilize around a certain value. Therefore two alternative outcomes are symmetric if their relative frequencies are the same in the long run. The relative frequency interpretation has some deficiencies with respect to causality. First, the requirement of infinite sequences of experiments means that probabilities of (causal) events which cannot even in principle occur infinitely often do not exist by this definition. To circumvent this, it has been suggested that infinite sequences are idealizations concerning what the limiting frequencies *would be* if the number of experiments would increase without bounds. Another unnatural property regarding causality is that the limiting frequency is symmetrical in the sense that the probabilities

$$P(E|C) = \frac{P(EC)}{P(C)} = \frac{P(C|E)P(E)}{P(C)}$$

and

$$P(C|E) = \frac{P(CE)}{P(E)} = \frac{P(E|C)P(C)}{P(E)}$$

are all equally well-defined. Of course, the temporal restriction that C should precede E can be used to rule out certain probabilities as meaningless, but formally the relation is symmetric. Moreover, causal dependence between singular events is a meaningful problem but cannot be evaluated in terms of relative frequencies.

A more recent interpretation of physical probability is the *propensity* approach (e.g. Popper, 1959, Hacking, 1965, Giere, 1973) which has considerable support among philosophers. The propensity interpretation was originally introduced by Popper (1957, 1959) to interpret individual events, not only sequences of trials as in the relative frequency approach.

Propensities have the interpretation of dispositions as causal tendencies; to bring about specific outcomes under relevant test conditions. These causal tendencies are properties of probabilistic experimental arrangements (chance setups), not events themselves. The analogy is with nonstatistical dispositions like fragility; a glass has the property to break when dropped, and it has this property whether it will actually be dropped or not. Propensities are therefore probabilistic causal conditionals which are incorporated with a simple bring about relation. They are theoretical concepts which are *causing* or producing empirical relative frequencies. Propensities, like physical probabilities in general, are assumed to be determined by physical laws, and therefore exist regardless of human knowledge. They are, by definition, asymmetrical unlike relative frequencies, and therefore more suitable to causal ideas.

Suppes (1987) remarks that propensities provide a certain causal view to probabilities which will qualify as a special kind of objective probability but propensities as such are not probabilities. This view is shared by Humphreys (1985) and Salmon (1988). Propensities do not, for example, fulfil Bayes' theorem. It is meaningless to talk about inverse propensities in the same spirit as we do about inverse probabilities. Salmon (1988) gives an example in which n machines produce floppy disks and the relative frequencies of defective products are known for each machine. Consider then a particular disk and the probability that it is defective. By the usual argumentation, using Bayes' theorem, we could calculate the probability that the particular disk was produced by a particular machine. However, it is not sensible to talk of the propensity of the disk to have been produced by a certain machine, even though it is perfectly sensible to say that a machine has a certain propensity to produce defective disks. In this sense, the propensities do not fulfil the rules of probability calculus. Instead, Salmon suggests that propensities are natural concepts for probabilistic causes. They produce the frequencies which he calls probabilities.

There seem to be considerable conceptual difficulties with the definition of propensities. Suppes (1987) finds it even impossible to present a general theory of propensities because 'propensity' is a general concept in the same sense as 'cause'; it is defined in the particular context in which it is used. Similarly, propensities are tendencies of something to act in a given manner under given circumstances, and these ingredients are different in each case.

1.4.2 Epistemic probabilities

We shall concentrate here on the subjective interpretation of epistemic probabilities. The distinction between objective and subjective probabilities seems to date back to the debate concerning the meaning of "equally probable". The question then raised whether this means that "according to perfect knowledge all the relevant circumstances are the same" or "because of ignorance of any difference in relevant circumstances" (principle of indifference). In the subjective probability interpretation the role of information is already built in the structure of the probability concept. Subjective probability is always conditional on the knowledge of the interpreter. Thus, it is a conditional probability although the condition is usually implicit. Subjective probabilistic statements are partial implications between the hypothesis (e.g. occurrence of an event) and relevant evidence. Since the probability is assumed to vary among individuals, subjective probabilities are also called *degrees of belief* or *personal* probabilities. They are understood behaviouristically; i.e., subjective probability expresses how a person would behave in a (hypothetical) situation. An individual's assertion about an event, which has not yet happened, may be certain or impossible, but there is a whole range of possibilities between these two extremes. These are the possible values of the subject's degree of belief concerning the event. Subjective probabilities can therefore be taken as reflections of the *uncertainty* of our knowledge. In subjective decision theories probabilities are rational degrees of belief in behavioristic sense.

Subjective probabilities must, in order to be proper probabilities (i.e. fulfil the axioms of probability), be *coherent*. The coherency principle is motivated by expressing subjective probabilities as betting quotients. In short, if the subject is willing to pay the sum pS to gain S in the case the event E is true (and lose pS if E is false), then p is the subject's degree of belief on $E's$ occurrence. The set of such betting quotients is called coherent if there no way to construct *a Dutch Book* against the subject. A Dutch Book is a set of bets which inevitably leads to loss whatever the outcome. Coherent betting quotients are shown to be probability measures (Ramsey, 1926, de Finetti, 1931). Some subjectivists, e.g. de Finetti (1968), consider the coherency principle to be everything a theory can prescribe. The choice among the coherent probability distributions should be left to the researcher.

Another difference between objective and subjective interpretations is the dynamics of the probabilities. Since subjective probabilities are conditional of the interpreter's state of knowledge, and are not assumed to exist objectively in nature like physical probabilities, they also change when the state of knowledge changes. In Bayesian learning theory the probabilities change as a consequence of new, additional information, real or hypothetical. When the new information H is assumed true, the prior probability $P(E)$ (or $P(E|A)$ if A is the initial state of knowledge) changes to a posterior probability $P(E|H)$ according to Bayes' theorem. The posterior probability of the event E given H is then

$$P(E|H) = \frac{P(H|E)P(E)}{P(H)} = \frac{P(H|E)P(E)}{P(H|E)P(E) + P(H|\bar{E})P(\bar{E})}$$

Some basic notions, like independence of random variables, are in Bayesian theory replaced by a weaker assumption of *exchangeability* which means that the random variables are symmetrical with respect to subjective probability considerations. In a sequence of trials it means that all permutations of the random variables have equal probability; i.e., their ordering plays no role. In his representation theorem de Finetti (1937) shows that in the case of exchangeable random variables, subjective probability measure can in fact be expressed as an expectation of an objective unknown probability measure (the limiting frequency) when the mixing distribution is the subjective probability distribution function of the objective probability. In particular, consider a sequence of exchangeable events $E_1, E_2, ...$. Let f_N be the number of events that occur. Then $\{f_N/N\}_{N\geq 1}$ converges (a.s.) to a random quantity θ whose (subjective) probability distribution function is indicated by F_θ. Then, for $x = 0, 1, 2, ..., N$ and $N = 1, 2, ...$ we have

$$P(f_N = x) = \binom{N}{x} \int_0^1 \theta^x (1-\theta)^{N-x} dF_\theta(\theta)$$

i.e., if θ is the unknown objective probability of independent binomial trials, then the subjective probability of x "successes" in N trials is a mixture of the objective probability weighted by (a priori) subjective assertion of its values on [0,1].

Conversely, fixing the values of the unknown θ, say, $\theta = \theta_0$ we will get a sequence of exchangeable events $E_1, ..., E_N$ by $P(E_i|\theta = \theta_0) = \theta_0$, $i = 1, ..., N$. This corresponds to assuming independence of random variables conditional on a parameter in Bayesian inference. The parameter θ represents in this case the limiting relative frequency of occurrence of the events. De Finetti argues that the notion of objective unknown probabilities without any reference to human evaluation is actually meaningless because all probabilities, according to his theorem, can be expressed as subjective probabilities. A less radical view is to consider it merely as a way to approximate the "true" objective probabilities.

Let us now return to the literature of causality. In the propensity approach the probability is actually interpreted in terms of causation, the bring about relation, whereas otherwise causation is interpreted in terms of probabilities. Roughly speaking, physical objective probabilities correspond to the idea of indeterministic causality, inherent randomness in nature in which non-deterministic causal laws determine the true objective probabilities. Then probabilities in causality relate to the ontological question of how causes actually operate. Subjective interpretation of probability in the Laplacean sense means that even though causal processes are in fact deterministic, our knowledge about them is always incomplete, and thus probabilities express the uncertainty of the causal statements. This is clearly an epistemic question. In general, subjective probabilities which change as a result of new knowledge reflect the belief that causal statements are actually always dependent on the interpreter's level of understanding of the underlying causal mechanisms.

The propensity interpretation is supported by Good (1988) who expresses it clearly: "Probabilistic causality is something that exists even if no conscious being is around". The propensities in his quantitative formulation are assumed to be given and conditional on "the state of the universe" and "existing physical laws". We already pointed out that even though Salmon (1988) does not give the propensities a status as proper probabilities, he evaluates them as probabilistic causes existing in nature. Subjective probabilities are in his view our best guesses of the strength of these probabilistic causes, and the role of "real" probabilities should be given to relative frequencies.

Davis (1988), in his criticism against probabilistic causal theories, also rules out subjective probability as a proper probability concept in causality since "...the existence of causal connections, unlike our knowledge about them, does not depend on our evidence". Davis concludes that since causation is not subjective nor relative to evidence, any plausible probabilistic theory of causality must define it in terms of empirical probability, that is, in terms of long-run relative frequencies. Holland (1986) also supports the relative frequency interpretation in causality: "A probability will mean [in Rubin's model] nothing more, or less, than a proportion of units in U [population]".

Suppes (1984) considers the question of the sense of probability meaningless. He writes: "Each of these cases [theoretical probability, relative frequencies, expression of belief] has its appropriate and proper place in the expressions of causal claims". Skyrms (1989) proposes that the subjectivist theory of conditional chance

(chance meaning physical probability) is the bridge between different interpretations of probability in causality (cf. de Finetti's representation theorem). If all that is relevant to the occurrence of an event is known, then the subjective probability (degree of belief) is equal to the chance, otherwise the degree of belief of an outcome is the expectation of its chance. The chances can then be recovered by conditioning on an appropriate partition, or more generally, on a σ-field, so that there is always a link between subjective and objective probabilities. The level of coarseness of the conditioning partition depends on the context of a particular causal analysis. It corresponds to the idea of controlling for the pre-existing conditions. Skyrms therefore classifies the theories of, for instance, Suppes and Granger (see Section 1.6.3) as subjective because "the raw statistics roughly plays the role of degree of belief, and probability conditional on the partition plays the role of "relevant" objective probability".

1.5 Counterfactuals in causality

We consider briefly the principles of counterfactuals since the idea of counterfactual comparison is central in some statistical approaches to causality. Counterfactual conditionals are of the type: "If a which is not F were F, then a would be G". In counterfactual conditionals the antecedent is known to be false. They thus provide a way to analyse particular non-actual but possible situations.

Two well-known theories of counterfactuals by Lewis (1973) and by Stalnaker (1984) use the concept of *possible worlds*. According to Stalnaker (1984) this notion can be understood as a stock of hypothetical beliefs. Consider a collection of possible worlds. Suppose that in each possible world every statement is either true or false. Assume that for the actual world w and for every possible condition A in w there exists a unique closest possible world w' in which A is true. The differences in w and w' are determined only by the condition A. The truth value of the entire counterfactual conditional "If A were the case then B would be the case" is entailed by determining whether B is true in the closest possible world w'. The conditional is true in the actual world w if B is true in the possible world w'. Lewis (1973) considers causal relations as counterfactual relations. In terms of counterfactuals a causal statement would be: "A causes B if A is true in the actual world and B is true in actual world, and in the closest possible world in which A is not true B is not true". In causal terms the same would be: "A causes B if and only if A occurs and B occurs and if A had not occurred then B would not have occurred".

In the philosophical literature the notion "similarity of possible worlds" has raised severe criticism. For example, Suppes (1984) finds it is unclear and difficult to define. In probabilistic terms it is, on the other hand, perfectly natural to talk about possible outcomes of events and construct a probability measure for these outcomes. Despite the criticism, the counterfactual principle seems to be able to capture an essential part of causal reasoning: the occurrence of a cause should have an influence which is different from what would have happened without the occurrence of it. In probabilistic terms, the probability of the effect (for a subject) should be different when actually observing the cause compared to the then hypothetical case in which the cause did not happen (for the same subject). The counterfactual principle in statistical analysis of causality has been followed at least in Rubin (e.g. 1974), in Holland (1986) and in Robins (1989).

1.6 Causality in statistical analysis

After this rather lengthy presentation of philosophical and probabilistic ideas in causality we now turn to statistical analysis in causation studies. We consider different ways of measuring and modelling causal dependence in some applied fields and in some, rather few, articles written by statisticians. Although it seems that statisticians have been rather reluctant to take any definite position in the relation between statistics and causality, there has quite recently been more discussion on it (e.g. Holland, 1986, Dempster, 1990, Rubin, 1990, 1991, Cox, 1992). In all these articles it is emphasized the fundamental role of causality also in statistical thinking. Dempster (1990) compresses the relation between causality and statistics into two common principles: the theory and practice of experimentation and the concept of replication. Experimentation implies control of treatment and confounding factors in order to be able to measure hypothesized causal effects. This principle concerns both randomized experiments and observational studies. On the other hand, a causal effect is something repeatable under certain conditions. This is what we assume in sample studies.

In what follows we classify the statistical approaches roughly in three groups: randomized experiments, association models and dynamic models. They are not entirely distinct but represent different traditions in empirical causal analysis. In conjunction with randomized experiments we shall briefly discuss some principles related to the vast literature of measuring causal effects in epidemiology in the presence of confounding. In general, the emphasis in the first group is on the observational plan and on the requirements that should be fulfilled in order to make causal inference from data. The mode of statistical inference is left aside. Some possibilities are considered in the review articles of Rubin (1990, 1991). The other two groups rely on regression methods, and they differ perhaps most in the way time is encountered in the models. We shall first present the basic ideas in each approach, and then compare them in the light of certain principles related to causality.

1.6.1 Randomized experiments

In randomized experiments the outcome of two or more treatments, which are assigned randomly to the experimental units, is compared under circumstances in which other factors are held fixed. In this setting the causal variables (causes and effects) are assumed known and the interesting causal question is the *relative* importance of the treatments (causes). In the simplest situation we have just one treatment and its complement, no treatment. As in traditional examples of randomized experiments (e.g. in agriculture), the outcome variable is not just an indicator of an event (as in event-causality) but rather a real-valued variable (e.g. the size of the yield of a particular treatment). The causal effect of a particular treatment is the difference between its outcome variable and other outcome variables.

Rubin (1974, 1978, 1990) and Holland (1986, 1988) consider randomized experiments along the line of Fisher (1935) to be the classical way of causal analysis in statistics. In contrast to what they call nonstatistical discussions of causality which are concerned on determining *the* cause of a particular outcome, the objective of randomized experiments is to learn about relative causal effects of various manipulations. Holland (1986) finds statements like "A is the cause of B" even false since they are always dependent on our current stage of knowledge of the subject matter, which is changing in time. Rubin (1990) considers that in a complex situation with many potential causes affecting the outcome a realistic task is to "describe the situation carefully to define the proposed treatment (cause) that has occurred and define which other potential causes are being assumed fixed at their observed values, and what *counterfactual alternative treatments* (causes) are being contemplated to have occurred rather than the observed treatment, which is the posited cause". The counterfactual approach means that a statistical study for causal effects always implies a comparison of results between at least two, and usually even more treatments on a population of units; between the one which was actually performed and the counterfactual other treatments. The simplest setting for this kind of comparison is a randomized experiment. Next we briefly characterize the structure of what is called a Rubin's model. We shall introduce it in the simplest situation of two treatments only.

The definition of Rubin's model requires the definition of units, treatments and outcome variable. Let $u \in U$ be individuals in population U, and variable S is defined on each u so that $S(u) = C$ if u received treatment C and $S(u) = \bar{C}$ if u did not receive treatment C. The outcome $Y_S(u)$ is a real-valued variable with $u \in U$ and $S = C$ or $\bar{C}$. In practice, we observe either $Y_C(u)$ or $Y_{\bar{C}}(u)$, depending on whether u received treatment C or not, even if the two values *could* have been observed.

The causal effect of the treatment C on Y is measured by the *individual* level causal parameter which is the difference of the outcome values

$$Y_C(u) - Y_{\bar{C}}(u). \quad (1.1)$$

Here $Y_{\bar{C}}(u)$ is unobserved if u received the treatment, and similarly, $Y_C(u)$ is unobserved if u belongs to the control group $\bar{C}$. This means that usually one cannot compute the individual level causal effect. The statistical solution is to compute the *average* or *population* level causal effect of the treatment over all $u \in U$, which is

$$E(Y_C - Y_{\bar{C}}) = E(Y_C) - E(Y_{\bar{C}}). \quad (1.2)$$

What can be estimated from the data (S, Y_S), however, is

$$E(Y_C|S=C) - E(Y_{\bar{C}}|S=\bar{C}). \quad (1.3)$$

The differences of the expectations in (1.2) and (1.3) are usually not equal. The equality depends on how the units were assigned to the treatment and control groups. If they were assigned *randomly* to each group then

$$E(Y_C) = E(Y_C|S=C)$$

and

$$E(Y_{\bar{C}}) = E(Y_{\bar{C}}|S=\bar{C}) \quad (1.4)$$

because then the treatment variable S is independent of the outcome variable Y. In this case the observed values of (S, Y_S) can be used to estimate the true average causal effect. In general the expectations are not equal because of *confounding* due to external factors.

The link between randomization, unbiasedness and (epidemiologic) confounding has been discussed, for example, by Greenland and Robins (1986). They suggest that confounding should be interpreted as *non-identifiability* of causal effects (or parameters in general) from the data. In order to be able to estimate causal effects from data we need additional comparability assumptions which relate to the Bayesian analogue of *exchangeability* between the comparison groups. Causal effects are estimable if it can be assumed that the proportion of response in each group is the same *if exposure is absent* (also Greenland, 1990). This is usually not true if there is confounding from external factors. Randomization is a means to decrease confounding but, as the authors emphasize, it does not prevent from it completely.

Holland (1986) gives a comparative review of different approaches to causality. His main criticism against the probabilistic causality theories (e.g. that of Suppes) is that causality is defined on aggregate level and not on individual level. He emphasizes that causal relations always exist on individual level even if one were interested in average level causal effects. In Holland's view Suppes' theory does not even possess the machinery to express the effect of a cause in a particular case since there are no units defined. Under the assumption of random assignment mechanism there is a correspondence to event-causality. If we let the outcome variable Y_S be the indicator of the *event* E, given treatment $S=C$ or $\bar{C}$, then

$$E(Y_C) = P(Y_C = 1), \text{ (resp. } E(Y_{\bar{C}}) = P(Y_{\bar{C}} = 1))$$

and

$$E(Y_C) - E(Y_{\bar{C}}) = P(Y_C = 1) - P(Y_{\bar{C}} = 1).$$

When using our previous notation, this is just

$$P(E|C) - P(E|\bar{C});$$

i.e., it is the difference of the same conditional probabilities Good (1961/1962) used when defining the causal tendency of C to produce E. Good also assumed an experimental setting in which case the probability of the cause, $P(C)$, should not affect the causal analysis. In Rubin's model the assumption of randomization is crucial since otherwise it would not be possible to estimate the true average causal effect from the observed data. In nonrandomized observational studies the independence assumption between the assignment mechanism and the outcome Y cannot usually be verified.

More generally, suppose we have the treated group C and the control group $\bar{C}$ with observed covariates $Z = z$ for all $u \in U$. Then the regression of Y_C on z in the treated sample, $E(Y_C|S = C, Z = z)$, and the regression of $Y_{\bar{C}}$ on z in the control group, $E(Y_{\bar{C}}|S = \bar{C}, Z = z)$, can be estimated directly from the data, and their difference is

$$E(Y_C|S = C, Z = z) - E(Y_{\bar{C}}|S = \bar{C}, Z = z). \qquad (1.5)$$

Again, this does not generally equal the average causal effect of C at z which is

$$E(Y_C|Z = z) - E(Y_{\bar{C}}|Z = z). \qquad (1.6)$$

If the assignment mechanism S to C or $\bar{C}$ is random then

$$P(S|Y_S, Z = z) = P(S|Z = z); \qquad (1.7)$$

i.e., S and Y are conditionally independent given the covariate values z. Then the observed prima facie causal effect equals the true average causal effect. Rosenbaum and Rubin (1983) call the treatment assignment *strongly ignorable* if, in addition to (1.7), the condition

$$0 < P(S|Z = z) < 1$$

holds for all z in the case of treatments C and $\bar{C}$ only.

We shall conclude this section by an example from epidemiology in which the counterfactual ideas in the same spirit as Rubin's and Holland are used to handle a confounding covariate in longitudinal setting. Robins (1989) considers the problem of finding an unbiased estimator for the true causal risk difference in the presence of a time-dependent confounder. He presents conditions under which the causal risk difference equals the crude risk difference which can be estimated from data. The values of the response and exposure status and the values of the confounding variable (as an example Robins considers death, exogenous oestrogen and blood cholesterol level) are recorded at times $t_1, ..., t_S$; t_S being the end of the study. Robins then proposes a nonparametric estimator called the *extended standardized risk difference* for the causal effect of the exposure when controlling for a time-dependent confounder. Although the notation of Robins is somewhat unconventional, we follow it in order to avoid misinterpretations.

Consider first a simple follow-up with three time points t_1, t_2 and t_3. Let $\mathcal{E} = \{(ee), (\bar{e}e), (e\bar{e}), (\bar{e}\bar{e})\}$ be the set of possible exposure histories at t_1 and t_2 (i.e., e=exposed, $\bar{e}$=unexposed). Then, as in Rubin's model, we consider four hypothetical outcome variables, the times of response (death) $T_{ee}, T_{\bar{e}e}, T_{e\bar{e}}$ and $T_{\bar{e}\bar{e}}$, from which only one can be observed for each individual. To be able to estimate the causal risk difference from data we need a set of assumptions concerning the comparability of the exposed and unexposed groups. If these two groups are exchangeable at $t_k, k = 1,2$, the probability of response when exposed and the "counterfactual" probability of response in the exposed *if they were unexposed*, are equal. Informally speaking this means that, in the absence of exposure, there are no external confounding factors which affect the response probability in the way that the two groups would differ in response. In the case of three time points this can be formalized as follows:

If the comparison groups are exchangeable at t_1 then (in terms of survival probabilities) the following holds

$$P(T_E > t_k|e(t_1)) = P(T_E > t_k|\bar{e}(t_1)), \quad k = 1,2 \quad (1.8)$$

for all hypothetical $E \in \mathcal{E} = \{(ee), (\bar{e}e), (e\bar{e}), (\bar{e}\bar{e})\}$, where $e(t_1)$ (resp. $\bar{e}(t_1)$) is the actual exposure status at t_1. For example, $P(T_{\bar{e}e} > t_k|e(t_1))$ is the unestimable "counterfactual" survival probability of those exposed at t_1 if they were unexposed at t_1. If (1.8) holds for all $E \in \mathcal{E}$ then the design is randomized through t_1 and the causal risk difference equals the estimable crude risk difference (cf. (1.4))

$$P(T_{\bar{e}\bar{e}} > t_2|\bar{e}(t_1)) - P(T_{ee} > t_2|e(t_1)). \quad (1.9)$$

If (1.8) holds and furthermore the comparison groups which have the same exposure at t_1 but differ at t_2 are exchangeable at t_2, the design is randomized through t_2. The following conditions are then needed

$$P(T_E > t_3|e(t_1), e(t_2)) = P(T_E > t_3|e(t_1), \bar{e}(t_2)), \quad (1.10)$$

for $E \in \mathcal{E}' = \{(ee), (e\bar{e})\}$, and

$$P(T_E > t_3|\bar{e}(t_1), e(t_2)) = P(T_E > t_3|\bar{e}(t_1), \bar{e}(t_2)), \quad (1.11)$$

for $E \in \mathcal{E}'' = \{(\bar{e}e), (\bar{e}\bar{e})\}$. If (1.8),(1.10) and (1.11) hold, the extended causal risk difference equals the estimable extended crude risk difference at t_3 which is

$$P(T_{\bar{e}\bar{e}} > t_2|\bar{e}(t_1))P(T_E > t_3|\bar{e}(t_1), \bar{e}(t_2)) - P(T_{ee} > t_2|e(t_1))P(T_E > t_3|e(t_1), e(t_2)) \quad (1.12)$$

The same reasoning can be extended to $S > 3$. The exchangeability condition corresponding to (1.8),(1.10) and (1.11) in the case of $S = 3$ is then

$$P(T_E > t_k|E(t_{r-1}), e(t_r)) = P(T_E > t_k|E(t_{r-1}), \bar{e}(t_r)), \quad 1 \leq r < k \leq S \quad (1.13)$$

where again $E \in \mathcal{E}$ denotes all possible exposure histories before k, the initial part of which up to $r-1$ is $E(t_{r-1})$ and $e(t_r)$ is exposure at t_r (resp. $\bar{e}(t_r)$ no exposure at t_r).

If the design is randomized through t_k the extended causal risk difference through t_k equals the extended crude risk difference through t_k which is

$$\prod_{m=2}^{k} P(T > t_m|\bar{E}(t_{m-1}), T > t_{m-1}) - \prod_{m=2}^{k} P(T > t_m|E(t_{m-1}), T > t_{m-1}) \quad (1.14)$$

where $E(t_{m-1})$ is exposure at all $t_m, m = 2, ..., k-1$ and $\bar{E}(t_{m-1})$ no exposure at any of $t_m, m = 2, ..., k-1$.

If we in addition control for a possible confounder L (blood cholesterol level), which is itself random and can also be an intervening variable for exposure and response, the condition for a randomized design through t_k is

$$P(T_E > t_k | E(t_{r-1}), L(t_r), e(t_r)) = P(T_E > t_k | E(t_{r-1}), L(t_r), \bar{e}(t_r)) \quad (1.15)$$

for all $t_r < t_k < t_S$, arbitrary $L(t_r)$ and all $E \in \mathcal{E}$, where $\mathcal{E}$ is the set of all possible exposure histories before t_k. Again $E(t_{r-1})$ is the initial exposure history of E until t_{r-1}, and $\bar{e}(t_r)$ is exposure at t_r. Then the extended *standardized* risk difference through t_k can be estimated from data as

$$\sum_{L(t_k)} \{ \prod_{m=2}^{k} P(T > t_m | L(t_{m-1}), T > t_{m-1}, \bar{E}(t_{m-1}))$$

$$\times \prod_{m=2}^{k-1} P(L(t_m) = l_m | L(t_{m-1}) = l, T > t_{m-1}, \bar{E}(t_{m-1})) P(L(t_1) = l_1) \}$$

$$- \sum_{L(t_k)} \{ \prod_{m=2}^{k} P(T > t_m | L(t_{m-1}), T > t_{m-1}, E(t_{m-1})) \quad (1.16)$$

$$\times \prod_{m=2}^{k-1} P(L(t_m) = l_m | L(t_{m-1}), T > t_{m-1}, E(t_{m-1})) P(L(t_1) = l_1) \}$$

by summing over all possible confounder level histories, $L(t_1)$ being the initial level at t_1.

If in addition the following two conditions concerning the confounder L hold, the extended standardized risk difference equals the extended crude risk difference through $t_s, s \leq S$ (for proofs, see Robins, 1989). First, the confounder is not an empirical risk factor for response when controlling for exposure; i.e., for any two confounder histories $L(t_r)$ and $L'(t_r)$, it holds that

$$P(T < t_k | E(t_{k-1}), L(t_{k-1})) = P(T < t_k | E(t_{k-1}), L'(t_{k-1})), \quad t_k \leq t_s \quad (1.17)$$

and second, the confounder is not a predictor of exposure. Then, for any two confounder histories $L(t_r)$ and $L'(t_r)$, it holds that

$$P(E(t_k) | E(t_{k-1}), L(t_{k-1}), T > t_k) = P(E(t_k) | E(t_{k-1}), L'(t_{k-1}), T > t_k), \quad t_k \leq t_s \quad (1.18)$$

Then (1.13), (1.15), (1.17) and (1.18) together imply that the standardized risk difference is the same as the estimable true risk difference.

Randomized experiments offer certain means to assess causal relationships which are seldom available in observational studies. Exchangeability of treated and non-treated units allows one to expect that there is no confounding due to external sources but this assumption is not easily fulfilled in observational studies. What is missing in Rubin's and Holland's analysis is the role of time. As Cox (1992) points out, the explicit notion of underlying causal processes is lacking. This has the consequence that it is not clear what is held fixed when varying the treatment variable. The role of intervening variables and their effect is not included in the simplified setting of Holland (1986), but elsewhere (Holland, 1989) he points out that intervening variables should be handled in the same manner as the outcome variable, being potentially affected by the treatment and thereby affecting the relationship between the treatment and the outcome. Robins (1989) extends the static setup of Rubin's model to longitudinal analysis where the exposure status and the values of a confounder or an intervening variable change in time. In this sense the ideas of Robins have much in common with those we shall present in Chapter 2.

1.6.2 Independence models

There is a long tradition of causal modelling in the social sciences. In *path analysis* (Wright, 1921) the association structure between a set of interval scaled variables is characterized by graphs and arrows. The *path coefficients* describing the strength of association between variables are calculated from the observed correlation matrix. Similar considerations of the relations between several variables in terms of partial regression coefficients date back already to Yule (1903).

Graphical models extend the ideas of path analysis to a more general framework (e.g. Kiiveri & Speed, 1982, Whittaker, 1990, for a review). These models, which are used in the structural analysis of multivariate data, are sometimes also called *independence models* since the dependence structure in the data can be expressed by independence constraints imposed on the variables. Several methods based on correlation analysis or regression techniques (e.g. factor analysis, path analysis, latent class models, log-linear models, structural equation models) can be formulated as *independence graphs*. The causal structure of the model is expressed by *marginal* and *conditional* indepedences between the variables. Both quantitative and qualitative models as well as latent variable models can be formulated in these terms.

We begin with some basic concepts and turn then to the graphical representation of causal models. Causal models with quantitative variables are presented as a system of regression equations (structural equation models); in the simplest case as linear equations. It is useful to borrow some concepts of econometrics here. We call variables that are explained by the model *endogenous*, and variables that are regarded as given or fixed by the theory *exogenous*. So, the model does not specify the random structure in the exogenous variables. In addition there are disturbance terms (errors or omitted causal variables) which are assumed to be uncorrelated with the exogenous variables, as usually in the linear models theory. The most general model would allow each endogenous variable to act as an effect and as a cause to other endogenous variables. In causal models we, however, require that a cause precedes the effect. In *recursive* models the possibility of reciprocal causality is ruled out. In a hypothetical example of a three-wave panel data the variable $Y(t_1)$ below (being also exogenous) can be a cause to $Y(t_3)$ but not vice versa.

$$Y(t_1) = e(t_1)$$
$$Y(t_2) = b_{21}Y(t_1) + b_{22}X(t_1) + e(t_2)$$
$$Y(t_3) = b_{31}Y(t_1) + b_{32}Y(t_2) + b_{33}X(t_1) + e(t_3)$$

The variable $X(t_1)$ is assumed exogenous and $e(t_i)$ an error term. *Block-recursive* models are recursive in blocks. They are formed of subsets of endogenous variables and associated disturbance terms. Causation is unidirectional between blocks and the disturbance terms are uncorrelated between blocks whereas within blocks there may be associations between exogenous variables and correlated disturbance terms are permitted.

To test hypotheses about causal dependence in the data, we need a set of predictions about the coefficients (e.g. Blalock, 1971). Unfortunately, the same set of predictions can yield models with a different causal interpretation. Without knowledge of the temporal sequence of the events, and other theoretical knowledge about their relationship, it is difficult to distinguish between different models. A principle, traditionally called *elaboration* in the sociological literature, is a procedure in which the relationship between two variables is tested in successive stages to find out whether it vanishes when other variables are controlled. The stages are often called explanation, interpretation and specification.

Consider two variables X and Y for which it is assumed that X causes Y ($X \rightarrow Y$). When the statistical significance of X on Y has been established, the aim of elaboration is to try to break down this relationship by first testing whether X actually is a *spurious* cause of Y in that there is a primary cause Z which causes both X and Y, and there is no true causal link between X and Y. This step corresponds to finding an *explanation* to $X \rightarrow Y$. As a consequence the regression coefficient of Y on X should vanish (subject to random variation). If this is not the case, the next step is to test whether an intervening cause I carries over the effect of X on Y so that it is actually *indirect*. Including I in the regression model of Y should again change the coefficient of X near to zero. This step of analysis is called the *interpretation* of $X \rightarrow Y$. It is easy to recognize the similarity of Suppes' principle of ruling out spurious causes from prima facie causes with the explanation step and Suppes' conditions for direct and indirect causes with the interpretation step. Finally, if $X \rightarrow Y$ still holds after the two previous steps (explanation and interpretation), the third step involves *specification* of the relationship. If the degree of dependence varies between the levels of some control variable C, an *interaction* term between X and C should be included in the model.

A simple example of spurious and indirect dependencies between variables X_1, X_2, X_3 and X_4 shows the logic of path analysis

$$\begin{array}{ccc} X_1 & \rightarrow & X_2 \\ \downarrow & & \downarrow \\ X_3 & \rightarrow & X_4 \end{array}$$

In this scheme no direct causal link exists between X_1 and X_4 or between X_2 and X_3. The recursive (one-way causation) equations for this model are

$$X_1 = e_1$$
$$X_2 = b_{21}X_1 + e_2$$
$$X_3 = b_{31.2}X_1 + e_3$$
$$X_4 = b_{42.13}X_2 + b_{43.12}X_3 + e_4 \tag{1.20}$$

If we control for the intervening variables X_2 and X_3 simultaneously, X_1 is not needed in the equation for X_4; i.e. the partial regression coefficient $b_{41.23}$ of X_4 on X_1, given X_2 and X_3, is restricted to 0. On the other hand, according to the model, the dependence on X_2 and X_3 is spurious because X_1 actually causes both, and controlling for X_1 means that $b_{32.1}$ is set to 0 in the equation for X_3.

The association structure in *graphical models* is given in terms of graph theory by vertices as variables, associations between them as edges in which case arrows are directed edges and lines undirected or symmetric edges. For example, the recursive models considered above are one special class of graphical models in that they are given by an oriented graph which is induced by complete ordering of the variables; i.e., there are no symmetrical associations (edges) between the endogenous variables. A graph is a *chain graph* if a *dependence chain* can be attached to it. A dependence chain is an ordered partitioning of the variable set into chain elements such that edges within them are undirected, and edges between them are directed pointing to the same direction. In order to be analyzable as graphical chain models, the chain elements must, in addition to the directional restriction of the edges, be ordered in a horizontal row (see Fig.1.1). A dependence chain determines the interpretation of each causal variable in the model. Different dependence chains (research hypotheses) may be presented by the same conditional independence structure in that they have the same underlying chain graph (same vertices and same edges). It is therefore important to note that the conditional independence structure determines the statistical structure of the model which is not unique with respect to different substantive research hypotheses.

The dependence chain determines how the *joint density* of all variables is obtained, and it also defines the conditional independence restriction for each variable pair which has a missing edge in the graph. For example, in Figure 1, which is Fig. 4 in Wermuth and Lauritzen (1990)

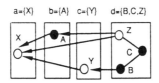

Fig.1.1 A graphical chain model with four chain elements a,b,c and d.

the joint density is $f_{abcd} = f_{a|bcd}f_{b|cd}f_{c|d}f_d$.

The joint distribution of the variables X_1, X_2, X_3 and X_4 in the example (1.20) is of the form

$$f(x_1, x_2, x_3, x_4) = f(x_1)f(x_2|x_1)f(x_3|x_1)f(x_4|x_2, x_3)$$

where X_2 and X_3 are the direct causes of X_4 but X_1 is not.

It is well known that in multinormal models linear independence constraints are particularly simple (e.g. Cox and Wermuth, 1993). *Marginal* independence between X_i and X_j implies

$$X_i \perp X_j \Leftrightarrow \sigma_{ij} = 0$$

and *conditional* independence can be expressed in terms of partial covariances or elements of the precision matrix Σ^{-1} as follows

$$X_i \perp X_j | X_k \Leftrightarrow \sigma_{ij.k} = \sigma^{ij} = 0$$

where $\{\sigma^{ij}\}$ is the (i,j)-element of Σ^{-1} and has the interpretation of the partial covariance of X_i and X_j when X_k has been regressed out. Conditional *directed* independence is given by the partial regression coefficient since

$$X_i \perp X_j | X_k \Leftrightarrow \beta_{ij.k} = \rho_{ij.k}(\sigma_{ii.k}/\sigma_{jj.k})^{1/2} = 0,$$

where X_k is usually a set of variables.

Qualitative recursive models are usually conjunctions of log-linear models (e.g. Kiiveri and Speed, 1982) and have therefore a log-linear parameterization. In this case, zero first (and higher) order interaction terms containing the pair (X_i, X_j) imply (conditional) independence of these variables. If (X_1, X_2, X_3, X_4) in the example (1.20) has a multinomial probability distribution; i.e., $P(X_1 = i, X_2 = j, X_3 = k, X_4 = l) = p_{ijkl}$, where $p_{ijkl} > 0, \sum_{ijkl} p_{ijkl} = 1$, the factorization of p_{ijkl} corresponding to the independence structure in (1.20) is $p_{ijkl} = p_{i+++}p_{i.j}p_{i.k}p_{ijk.l}$, where p_{i+++} is the marginal probability distribution of X_1, $p_{i.j}$ the conditional distribution of X_2 given X_1, and $p_{i.k}$ and $p_{ijk.l}$ are defined respectively.

Lauritzen and Wermuth (1989, 1990) have proposed a class of *mixed* association models where quantitative and qualitative variables can be considered in the same model. The joint distributions are then conditional Gaussian distributions of the continuous variables, given the discrete variables. We demonstrate the association between parameterization and conditional independence in more detail in the mixed models.

Following Lauritzen and Wermuth (1989), we let $V = \Delta \vee \Gamma$ be a set of variables partitioned to discrete, Δ, and continuous, Γ, variables with corresponding random variables denoted by $X_\alpha, \alpha \in \Delta \vee \Gamma$. A class of probability distributions expressed by the discrete, linear and quadratic canonical characteristics in the following form

$$\begin{aligned}f(x) = f(i,y) &= exp\{g(x_\Delta) + h(x_\Delta)'x_\Gamma - \frac{1}{2}x_\Gamma' K(x_\Delta)x_\Gamma\} \\ &= exp\{g(i) + h(i)'y - \frac{1}{2}y'K(i)y\}\end{aligned} \quad (1.21)$$

where g is a real-valued function of i, h a q-vector-valued function and K is $(q \times q)$-matrix-valued function of i, have conditional Gaussian distributions in the sense that x_Γ for given $x_\Delta = i$ is a q-variate Gaussian with covariance $K(i)^{-1}$ and expectation $K(i)^{-1} h(i)$; i.e.,

$$L(x_\Gamma | x_\Delta = i) = N_q(K(i)^{-1} h(i), K(i)^{-1}).$$

In Lauritzen and Wermuth (1989) several results are given how these distributions behave under conditioning and marginalization, and the form of the likelihood functions is given under different model classes.

As edges correspond to direct relations between the (causal) variables, indirect relations (missing edges) between the variables can be analysed in CG-models by their parameterizations with interactions. As noted earlier, a variable pair is conditionally independent given the remaining variables if all interaction terms containing this pair are zero. Estimates of all interaction terms corresponding to postulated missing edges are then close to zero. The parameterization with interactions is illustrated in the case $V = \{A, B, X, Y\}$ in which $\Delta = \{A, B\}$ and $\Gamma = \{X, Y\}$ with A and B having categories $i = 1, 2, ..., I$ and $j = 1, 2, ..., J$ and $X = x$ and $Y = y$ (Wermuth & Lauritzen, 1990). Then

$$log f(i, j, x, y) = g_{ij} + h^x_{ij} x + h^y_{ij} y - \frac{1}{2} k^x_{ij} x^2 - \frac{1}{2} k^y_{ij} y^2 - k^{xy}_{ij} xy \qquad (1.22)$$

which is in terms of interactions

$$\begin{aligned} log f(i, j, x, y) =& \lambda + (\lambda^A_i + \lambda^B_j + \lambda^{AB}_{ij}) + (\eta^X + \eta^{AX}_i + \eta^{BX}_j + \eta^{ABX}_{ij}) x \\ &+ (\eta^Y + \eta^{AY}_i + \eta^{BY}_j + \eta^{ABY}_{ij}) y \\ &- \frac{1}{2}(\phi^X + \phi^{AX}_i + \phi^{BX}_j + \phi^{ABX}_{ij}) x^2 \\ &- \frac{1}{2}(\phi^Y + \phi^{AY}_i + \phi^{BY}_j + \phi^{ABY}_{ij}) y^2 \\ &- (\phi^{XY} + \phi^{AXY}_i + \phi^{BXY}_j + \phi^{ABXY}_{ij}) xy \end{aligned} \qquad (1.23)$$

with suitable constraints on the summations of the parameters. Then, for example, conditional independence between A and Y, given B and X, means that the estimates of $\eta^{ABY}_{ij}, \phi^{ABY}_{ij}$ and ϕ^{ABXY}_{ij} are near zero. When restricting to discrete variables ($x = y = 0$) only, we get the ordinary log-linear model parameterization in a two-way contingency table.

Several authors have proposed automatic search procedures for graphical models to find the most likely causal structure of a data set (e.g. Kreiner, 1986, Verma and Pearl, 1990, Spirtes et al., 1993). These procedures start usually with undirected graphs, proceed by testing conditional independence of all possible pairs of variables given a set of other variables in the model, then moving to triples of variables and testing their orientation until no changes are needed anymore. No preliminary ordering of the variables is necessary which means that such analysis seems, more or less, exploratory.

The methods considered in this section rely upon certain sometimes rather restricting statistical assumptions such as multinormality and linearity. The starting point of the analysis is the correlation or covariance matrix. The principle of elaboration relates directly to causal inference but the usefulness of these models depends on how well the underlying causal mechanisms are understood. The models themselves are general association models, and they lack any specification of *causal* dependence. Dempster (1990) criticizes path analysis from its failure to start with conceptions of causal processes that relate directly and explicitly to constructed mathematical formulations. Obviously, the cross-sectional setting of the models is one reason for this. In addition, the causal concepts, the path coefficients, are completely model-dependent in the sense that any change in the model changes the interpretation and position of a coefficient relative to others. This is certainly not a satisfactory feature when trying to establish causal effects. The strength of these models is the possibility to represent underlying causal structures in explicit forms (e.g. graphical chains) and the way causal hypotheses can be handled by statistical modelling principles, for example, conditional independence in terms of interactions.

1.6.3 Dynamic models

In the two previous approaches the ordering of causal events is expressed only by the requirement that a cause must precede the effect. Even though Suppes initially considered the events as outcomes of stochastic processes, the exact occurrence times did not play any further role in his theory. In the following the assumption of time homogeneity of causal dependence is relaxed. The approach is dynamic in the sense that the present is predicted from the past observations. Typically then, the observations are considered as realisations of stochastic processes. In econometrics the so called Granger-causality (Granger, 1969, already in Wiener, 1956) is used almost routinely to analyse and test the dependence of two time series. According to Granger's definition, two time series $\{X_t,\ t = ..., -1, 0, 1...\}$ and $\{Y_t,\ t = ..., -1, 0, 1, ..., \}$ do not show causal dependence if the prediction of Y_{t+1}, given the pre-t history $\{..., Y_{t-1}, Y_t\}$, is not improved (in the mean squared error-sense) by including in the regression model the history $\{..., X_{t-1},\ X_t\}$; i.e., for all t

$$E(Y_{t+1}|Y_t, Y_{t-1}, ..., X_t, X_{t-1}, ...) = E(Y_{t+1}|Y_t, Y_{t-1}, ...). \quad (1.24)$$

Obviously comparing the conditional means only is a rather strong restriction. Chamberlain (1982) and Florens and Mouchart (1982, 1985) have generalized Granger's restriction on linear predictors to nonlinear case by comparing the entire conditional distributions. The definition for noncausality between X and Y is then

$$P(Y_{t+1} \leq y|Y_t, Y_{t-1}, ..., X_t, X_{t-1}, ...) = P(Y_{t+1} \leq y|Y_t, Y_{t-1}, ...), \text{ for all } t. \quad (1.25)$$

In particular, Chamberlain considers the situation when the response process Y is discrete-valued, and in the simplest case has only values 0 and 1 corresponding to event-causality. When Granger-causality is defined in terms of conditional independence it has obvious similarities with Suppes' theory if the condition is the whole history of the cause.

The same idea is present in Mykland (1986) when defining the concept of causality in filtered spaces for continuous time processes, and in Aalen's (1987) conditions for two random processes $X = (X_t)$ and $Y = (Y_t)$ to be *locally independent*. They put the ideas into the framework of semimartingales. Let $(\Omega, \mathcal{F}, P)$ be a probability space and $\mathcal{F}_t = \mathcal{F}_t^X \vee \mathcal{F}_t^Y$ the smallest history containing both $\mathcal{F}_t^X$ and $\mathcal{F}_t^Y$, where $(\mathcal{F}_t^X)$ (resp. $(\mathcal{F}_t^Y)$) is generated by X (resp. Y). Denote $(\mathcal{G}_t) \leq (\mathcal{F}_t)$ if $\mathcal{G}_t \subseteq \mathcal{F}_t$. Then X_t is $(\mathcal{F}_t)$-adapted if $(\mathcal{F}_t^X) \leq (\mathcal{F}_t)$. Mykland defines that $(\mathcal{F}_t^X)$ *entirely causes* $(\mathcal{F}_t^Y)$ *within* $(\mathcal{F}_t)$ *relative to* P, and denotes it by

$$(\mathcal{F}_t^Y) < (\mathcal{F}_t^X); (\mathcal{F}_t); P \qquad (1.26)$$

if $(\mathcal{F}_t^Y)$ and $(\mathcal{F}_t^X)$ are subfiltrations of $(\bar{\mathcal{F}}_t)$ (the completed filtration) and

$$P(A|\mathcal{F}_t^X) = P(A|\mathcal{F}_t) \qquad (1.27)$$

holds for all t and all $A \in \mathcal{F}_\infty^Y$. This corresponds to the conditional independence of $\mathcal{F}_\infty^Y$ of $\mathcal{F}_t$, given $\mathcal{F}_t^X$, for all t. Furthermore, (X_t) or $(\mathcal{F}_t^X)$ is *its own cause* (within $(\mathcal{F}_t)$ relative to P) if

$$P(B|\mathcal{F}_t^X) = P(B|\mathcal{F}_t) \qquad (1.28)$$

holds for all t and all $B \in \mathcal{F}_\infty^X$. Then (X_t) is called a self-exciting process.

Assume now that X and Y are two processes counting occurrences of events in time. A Doob-Meyer decomposition for X and Y w.r.t. $(\mathcal{F}_t)$ (e.g. Brèmaud, 1981) is of the form

$$X_t = \int_0^t \lambda_s^X ds + M_t^X, \quad Y_t = \int_0^t \lambda_s^Y ds + M_t^Y \qquad (1.29)$$

in which $\int_0^t \lambda_s^X ds$ is the systematic prediction (the compensator) of X (resp. Y) and the martingale M_t^X is the error in the prediction of X (resp. Y). Aalen (1987) calls the process Y to be locally independent of the process X over a time interval I if the two processes do not have simultaneous jumps and the stochastic intensity λ_t^Y of the process Y with respect to $(\mathcal{F}_t)$ is $(\mathcal{F}_t^Y)$-adapted; i.e., in a "small interval dt" the probability of a new event in the Y-process is $\lambda_t^Y dt = P(dY_t = 1|\mathcal{F}_t) = P(dY_t = 1|\mathcal{F}_t^Y)$, so that it depends only on Y's own history (cf. (1.25)). The definition of local independence was originally given by Schweder (1971) in the context of Markov chains. Aalen prefers the word local dependence to causality since inclusion of other processes than just X and Y can change the local dependence of all processes. The concept of local dependence is therefore completely dependent on the information one has and uses in a particular situation.

The main criticism against Granger-causality and its extensions of the form (1.25) is that causal dependence is analysed in aggregates as usually in time-series analysis. This means effectively, as Holland (1986) puts it, that variables cause other variables. The same problem concerns, of course, the association models in Section 1.6.2. A question like "Is crime causing poverty or poverty causing crime", which Granger (1986) himself raised in his comment to Holland, is perhaps not properly analysed by data on series of average crime rates and per capita income although the main interest is to understand changes in the average rates. Causal dependence concerns the inviduals committing crimes or being poor, and changes in their life-histories. Even though we may think that changes in legistlation or other relevant social reforms in this particular context affect all individuals at the same time, their influence can still be different on different individuals, and this may change the causal dependence one wants to measure. It has recently been recognized (e.g. Tuma & Hannan, 1984) that such complex causal dependencies should be analysed dynamically in terms of *individual event histories*. This approach is adopted in the statistical models of stochastic intensities in event-history analysis, and it will be our choice also in Chapter 2.

1.7 Discussion

The purpose of this chapter was to elucidate the fundamental role of causality in science and in particular, in statistical analysis. As mentioned in Section 1.3, the role of probability in causality corresponds to, on the one hand, the idea of inherent indeterminism in nature and, on the other hand, incomplete understanding or uncertainty of (perhaps deterministic) causal relations. A simplified interpretation is that probabilistic causality is just a weakening of strict causality which assumes invariable dependence between cause and response. Different interpretations of the notion of probability support these views in different ways. The propensity approach corresponds most closely to the idea of law like causality in the natural sciences where causal processes are much better known and causal propositions more easily experimented than in human studies or in the social sciences. The interpretation of the propensities as causal tendencies related to the setup, not to the outcomes, corresponds to the idea that specific circumstances ("test" conditions) have a tendency to bring about certain outcomes. In areas where statistical modelling is nowadays widely used (such as the social sciences or epidemiology) we are inevitably faced with the fact that our current knowledge of the causal processes is imperfect, and even the experimentation of causal propositions is sometimes impossible. Statistical modelling is therefore always more or less exploratory and reflects the researchers opinion and knowledge of the subject matter. Thus, the conditional probabilities obtained from such models are always subjective. For practical purposes, these considerations seem, however, rather academic.

In statistics causal dependence is often mixed with pure association between variables. In the following we try to distinguish certain principles which are connected to causality but not necessarily to association. A common view in the literature of causality is that a cause brings about a *change* in its effect, or in event-causality, the effect occurs *because* the cause is present. In deterministic causality the effect follows the cause with probability 1. In probabilistic causality the occurrence of a cause raises the probability or prediction of the effect from a level it has been or would have stayed without the cause. Measuring the effect of a cause therefore always implies comparing the situation with and without the occurrence of the cause. This obvious principle is present in different ways in all three approaches considered. In Rubin's model change means the difference in the effect of the actual treatment on the response compared to that of the other possible counterfactual treatments. In association models change means the amount of change in the dependent variable due to a unit change in a particular independent variable, and it is expressed by the path coefficient. In dynamic models change means change in the prediction of the response from t to $t+1$ due to adding to the conditioning history before t the information concerning the state of some relevant variables and the response itself at t.

Furthermore, causal dependence is a concept on individual or unit level; it is not the relative frequency of the cause in a population that brings about certain effects, it is the fact that the cause is present in certain individuals and brings about the effect in them. The conditional probabilities, given the cause and without it, should be compared on the same unit. Averaging can easily distort the causal effect in a given causal field if there is confounding due to some external uncontrolled factors. The construction of Rubin's model is explicitly based on unit level causality, and it is also explicitly stated that the estimability of causal dependence on aggregate level depends on the observational plan. In the other two approaches causal dependence is a relationship between variables, and often the unit level is not even defined.

A causal effect evaluated at a certain time t may not be the same as that at $t+k$ since the external circumstances have changed. It seems unrealistic to expect that one could isolate a dependence relation between a cause and its effect from these changing external circumstances except in simplified controlled experiments. In observational studies this is hardly ever possible. Sometimes also the lag between the occurrence of a cause and its effect, or the "response time", is of explicit interest. It seems therefore natural that causal analysis should be dynamic. The idea of controlling for *changes* in time seems to be implicit also in the criticism against the probabilistic theories and statistical approaches presented above. Salmon (1980) criticized the theories of Good and Suppes for not considering causal dependence in terms of causal processes evolving and intersecting in time, and Cox (1992) criticized Rubin's model for its inability to state clearly what is held fixed and what is allowed to change. Studies with association models are usually cross-sectional, or at best two- or three-wave panel designs with arbitrary time intervals, which easily fail to reveal the functioning of underlying causal processes. In contrast to that, in dynamic models the concept of change and learning from the past are essential.

It also seems desirable that the basic causal concepts, like causal effects, are expressed in probabilistic terms which are model-independent unlike regression coefficients of a particular statistical model. The interpretation and mutual position of these coefficients change as the model is respecified by transformations or new variables are included. The basic causal concepts (probabilities) are of course *estimated* from the data by assuming a certain statistical model which describes the substantive research hypothesis in a particular case. In Rubin's model the causal concept is the individual level difference $Y_C(u) - Y_{\bar{C}}(u)$ of the response. Since this difference is unobservable, the average or population level measure $E(Y|C) - E(Y|\bar{C})$ estimates (unbiasedly) the causal risk difference in the absence of confounding. In dynamic models the causal concepts are in the simplest case the differences $E(Y_{t+1}|Y_t, Y_{t-1}, ..., X_t, X_{t-1}, ...) - E(Y_{t+1}|Y_t, Y_{t-1}, ...)$, or more generally, the differences in the conditional probabilities, given the entire past history. In event-causality (when Y is just an indicator of the response) these two are equal. It seems that the most unsatisfactory way of presenting estimated causal dependencies is the direct use of regression coefficients as in the association models.

It is obvious that the three statistical approaches satisfy the above principles in various ways. In the next chapter we shall introduce a method for analysing causal effects in event-causality which has ingredients from all the three approaches considered, but since dynamics and learning from the past are its essential characteristics, it mostly resembles the dynamic approach. The classical concepts of Hume's causality: constant conjunction, succession in time and locality, can easily be replaced by probabilistic terms and be incorporated in statistical models. We just assume probabilistic dependence between a cause and its effect, and the requirement of a cause preceding the effect can be verified in a natural way in full event-history data. The last requirement, locality, corresponds in dynamic models to the Markov assumption. This assumption is not always realistic (e.g. Suppes 1986, 1990 and Granger, 1986) in causal chains, and we shall relax it.

2. PREDICTIVE CAUSAL INFERENCE IN A SERIES OF EVENTS

2.1 Introduction

Consider a sequence of k events $C_1, C_2, ..., C_k$, and denote the vector of their occurrence times by $\mathbf{S} = (S_1, ..., S_k)$. Let each event be connected to other events in the sequence either directly or by intervening events as in Figure 2.1. Assume that the sequence of events is a *causal chain* and let $C_k = E$ be the response or the "ultimate effect" (E) in the chain. Assume furthermore, that the chain is *ordered*; i.e., for the occurrence times holds $S_{r-1} < S_r$ whenever $S_{r-1} < \infty, r = 2, ..., k$. Then the chain is "forward going". By the notation $C \to E$ we shall throughout mean "C causes E" in the probabilistic sense if not otherwise mentioned.

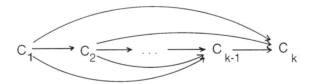

Fig. 2.1 An ordered k-point causal chain

In the preceding chapter an event C was defined to be a probabilistic cause of another event E if its occurrence raises the probability of E. A direct implication in a causal chain is that this holds at least for the adjacent events C_{r-1} and $C_r, r = 2, ..., k$. Whether it should hold also for "remote" links, such as $C_1 \to C_k$, has been discussed to some extent in the literature.

Consider two types of *causal transmission* in a chain of events which we illustrate in the following way:

$$C_1 \longrightarrow C_2, C_2 \longrightarrow C_3, ..., C_{k-1} \longrightarrow C_k \qquad (2.1)$$

and

$$C_1 \longrightarrow C_2, \ C_1 \cup C_2 \longrightarrow C_3, ..., C_1 \cup C_2 \cup ... \cup C_{k-1} \longrightarrow C_k, \qquad (2.2)$$

where $C_1 \cup C_2$ means "C_1 and C_2". They are extreme in the sense that in (2.1) only the immediate predecessor (i.e., C_{r-1} to C_r) is needed to transmit the causal influence whereas in (2.2) all preceding events are needed for causal transmission. In terms of deterministic causality, in (2.1) C_1 brings about C_2 which (independently of C_1 if there are other causes which can also bring about C_2) brings about C_3 etc. In terms of probabilistic causality we have, for example, $P(C_3|C_1C_2) = P(C_3|C_2) > P(C_3|\bar{C}_2)$; i.e., a chain with such a truncated dependence structure is a (first order) Markov chain. In (2.2) causal transmission is cumulative in that both C_1 and C_2 are needed to bring about C_3, and ultimately all preceding events $C_1, C_2, ..., C_{k-1}$ are needed to bring about $C_k = E$. In terms of probabilistic causality the occurrences of preceding events reinforce the influence of C_{r-1} on C_r so that, for example, $P(C_3|C_2C_1) > P(C_3|C_2\bar{C}_1)$. Obviously such causal chains are not Markovian. Between these two extremes there are many other possibilities which are illustrated in Fig. 2.1.

In the philosophical literature the form of causal transmission has been dealt with at least in Good (1961/1962), Salmon (1984) and in Suppes (1986,1990). Good (1961/1962) defines a causal chain to be a Markov chain in the sense of (2.1) where subsequent events have positive tendency to cause the next event. Salmon (1984) allows the influence of remote causes $C_1, C_2, ..., C_{k-2}$ be even negative on $C_k = E$ as long as each event in the series shows *positive statistical relevance* for its immediate successor in the chain. Suppes (1986, 1990) considers hypothetical causal chains of type (2.1) unrealistic in many practical examples. Non-Markovian causal chains are in other fields than in the physical sciences quite typical, and he finds it rather unlikely that, for example, in human studies or in the social sciences the "present" can be described in such detail that it would be sufficient for predicting future events.

In probabilistic causality the conditions for causal *transitivity* can be formalized in the case of a simple 3-point chain $C \rightarrow F \rightarrow E$ as follows (Suppes, 1986, Eells & Sober, 1983)

Theorem 2.1. If

(i) $P(F|C) > P(F|\bar{C})$

(ii) $P(E|F) > P(E|\bar{F})$

(iii) $P(E|CF) \geq P(E|\bar{C}F)$ and $P(E|C\bar{F}) \geq P(E|\bar{C}\bar{F})$

then $P(E|C) > P(E|\bar{C})$.

Proof. Using the fact that $P(E|C) = P(E|C\bar{F})P(\bar{F}) + P(E|CF)P(F)$ and conditions (i)-(iii) gives the result.

If the equality holds in both parts of (iii) the chain is Markovian and the influence of C is transmitted completely via F. The condition (iii) implies that C increases the probability of E regardless of the occurrence of the intervening event F (but the influence can still be stronger if also F occurs).

The joint influence of causes has also been discussed in some applied fields, for example, in epidemiology by Rothman *et al.* (1979) and Miettinen (1982).

Miettinen (1982) considers the interdependence between two symmetrical causes C_1 and C_2 in terms of the notions *synergism* and *antagonism*. In deterministic setting two cooperating causes show synergism if they are both needed to bring about the response E, and antagonism if the presence of either one empties the other's effect on E. In terms of probabilistic causality the causes are synergetic if $P(E|C_1C_2) \geq P(E|C_1\bar{C}_2)$ (or $\geq P(E|\bar{C}_1C_2)$, and both $P(E|C_1\bar{C}_2)$ and $\geq P(E|\bar{C}_1C_2)$ should then be ≈ 0. This is not the same as statistical *interaction* which can be formalized in terms of probabilities as $P(E|C_1C_2) \geq P(E|C_1\bar{C}_2)$ or $(\geq P(E|\bar{C}_1C_2))$; i.e., both causes add each other's influence on E but nothing is assumed about the size of their "main" effect on E, using the terminology of statistical modelling. Antagonism can be defined similarly in an obvious way. In applied fields, such as epidemiology, it is useful to make a distinction between *causative* and *preventive* causes on the response (e.g. Miettinen, 1982, Greenland & Robins, 1986). A typical example is a setup where some disease (A) increases the risk of death (C), whereas an intervening treatment (B) prevents from death. A natural question is then "What is the preventive effect of the treatment on death, given that a patient has contracted the disease?" In probabilistic sense the intervening treatment then "causes" the survival of a patient; i.e., raises the probability of survival.

In the next two sections we introduce a mathematical framework in which these rather vague probabilistic ideas of causal chains can be interpreted explicitly and in a dynamic context. We shall take a fairly liberal position when using the notion of a *causal* chain in conjunction with a series of events. We do not restrict the form of influence to be purely causative (positive). On the contrary, we think that in many interesting practical situations the influence can be either causative or preventive depending on how we define the response (E or $\bar{E}$). Instead of causal chains, one could of course use the word association chains, but since we consider strictly directional associations, we prefer the word causal. We also let the dependence on the preceding events be completely general between the extremes of (2.1) and (2.2). The appropriate dependence structure in a particular causal chain can then be given by defining, if necessary, some of the routes in Fig 2.1 as impossible.

2.2 The mathematical framework: marked point processes

A useful mathematical formulation of a causal chain can be given by a *marked point process*. The occurrence of the causal events $C_r, r = 1, .., k$, is then described by pairs of random variables (T_n, X_n), where $0 \leq T_1 \leq T_2 \leq ...$ are ordered time variables for which the equality holds only when the occurrence times are infinite. The marks $X_n \in E$ are descriptions of the causal events occurring at T_n. We denote by $(T, X) = \{(T_n, X_n); n \geq 1\}$ the marked point process.

There is a one-to-one correspondence between the occurrence times $S_1, S_2, ..., S_k$ of events C_r, and the point process (T, X) in the following way

$$T_1 = \inf_{1 \leq r \leq k} S_r;\; X_1 = \{1 \leq r \leq k; S_r = T_1\}$$

$$T_{n+1} = \inf\{S_r : 1 \leq r \leq k, S_r > T_n\};\; X_{n+1} = \{1 \leq r \leq k; S_r = T_{n+1}\}.$$

If all events occur and the chain is ordered, then obviously $T_r = S_r, r = 1, ...k$. Here we prefer, however, another way to describe the evolution of the process (T, X). For each $x \in E$, define the counting process

$$N_t(x) = \sum_{n \geq 1} 1_{\{T_n \leq t, X_n = x\}},\; t \geq 0. \tag{2.3}$$

which is a step function with right-continuous sample paths. If event x occurs at t, the process $N_t(x)$ counts 1 at time t.

We shall consider the marked point process (T, X) on the *canonical* space Ω in which case the elements of Ω are simply $\omega = (t_n, x_n)_{n \geq 1}$ so that $T_n(\omega) = t_n$ and $X_n(\omega) = x_n$. We will also use a "backwards" definition of the point process (T, X), called a *history process* by Norros (1986). It is formed of the marked points before t; i.e.,

$$H_t = \{(T_n, X_n); T_n \leq t\}. \tag{2.4}$$

The history process takes values in $(\mathbb{H}, \mathcal{H})$ where $\mathbb{H}$ is a subset of $(0, \infty] \times E$, the set of possible values of $(T_n, X_n)_{n \geq 1}$, and $\mathcal{H}$ are the corresponding Borel sets. The conditional probabilities used later on are conditioned on the increasing σ-fields generated by the history process; i.e., $\mathcal{F}_t = \sigma(H_t) = \mathcal{F}_t^N$, so that we simply condition on the *internal* history of the marked point process. Therefore, for all probabilities considered here it holds that $P(\cdot|\mathcal{F}_{T_n}) = P(\cdot|H_{T_n})$. The "usual conditions" concerning right-continuity and completeness of the filtration $(\mathcal{F}_t)_{t \geq 0}$ are assumed here.

The *compensator* $(\Lambda_t(x))$ of the counting process $(N_t(x))$ is a predictable process which admits the property that $N_t(x) - \Lambda_t(x)$ is a local martingale with respect to the filtration $(\mathcal{F}_t)_{t \geq 0}$ to which $(N_t(x))$ is adapted. This property is only mentioned here and is used in the next section, but otherwise we refer to the basic references of the general theory of point processes (e.g. Brémaud, 1981). Since we present here only the necessary mathematical background for our purpose, we refer in the case of marked point processes to Arjas (1989).

The way the probability distribution of the marked point process (T, X) is specified is particularly useful in analysing causal chains. We can specify the distribution point by point (i.e., event by event), starting from the probability of (T_1, X_1), given H_0, until the distribution of (T_{n+1}, X_{n+1}), given H_{T_n}. Although we will use hazards to specify distributions, we first define the "subdistributions" of certain marked points, since this notion is needed in Chapter 4. The distribution of (T_n, X_n), given $H_{T_{n-1}}$, is for $t > T_{n-1}$

$$P(T_n \leq t, X_n = x|\mathcal{F}_{T_{n-1}}) = F^{(n)}(t, x|H_{T_{n-1}}) \tag{2.5}$$

and the conditional distribution of (T_n); i.e., the probability of the n'th point before t, regardless of the mark, given $H_{T_{n-1}}$, is for $t > T_{n-1}$

$$P(T_n \leq t|\mathcal{F}_{T_{n-1}}) = F^{(n)}(t|H_{T_{n-1}}) = \sum_{x \in E} F^{(n)}(t, x|H_{T_{n-1}}). \quad (2.6)$$

The x-specific hazard of point (T_n, X_n) can now be given in terms of these by

$$d\Lambda^{(n)}(t, x|H_{T_{n-1}}) = \frac{dF^{(n)}(t, x|H_{T_{n-1}})}{1 - F^{(n)}(t-|H_{T_{n-1}})}, \ t > T_{n-1}. \quad (2.7)$$

Analogously with (2.6), the *crude* hazard of point (T_n) is the sum of the x-specific hazards

$$d\Lambda^{(n)}(t|H_{T_{n-1}}) = \sum_{x \in E} d\Lambda^{(n)}(t, x|H_{T_{n-1}}), \ t > T_{n-1}. \quad (2.8)$$

Conversely, when the conditioning history is $(\mathcal{F}_t^N)$, the hazards define the probability distribution uniquely. In the general case, where the compensators can be discontinuous, we have for $t > T_{n-1}$

$$dF^{(n)}(t, x|H_{T_{n-1}}) =$$
$$d\Lambda^{(n)}(t, x|H_{T_{n-1}})exp(-\Lambda^c(t|H_{T_{n-1}})) \prod_{T_{n-1} < s < t} (1 - \Delta\Lambda^{(n)}(s|H_{T_{n-1}})), \quad (2.9)$$

where $\Lambda^c(t|H_{T_{n-1}})$ in the exponential term is the continuous part of $\Lambda^{(n)}(t|H_{T_{n-1}})$. In the continuous case the last term on the right side is 1 since the jumps of the compensators, $\Delta\Lambda^{(n)}(t, x|H_{T_{n-1}})$, are 0. This formula shows also how the hazards define the likelihood completely when $\mathcal{F}_t = \mathcal{F}_t^N$.

Since the hazard $d\Lambda^{(n)}(t, x|H_{T_{n-1}})$ (and $F^{(n)}(t, x|H_{T_{n-1}})$) of the point (T_n, X_n) is conditioned on the history at T_{n-1}, it has support only in the interval $(T_{n-1}, T_n]$. Therefore the (x-specific) hazard is the sum of the local hazards

$$d\Lambda_t(x) = \sum_{n \geq 1} d\Lambda^{(n)}(t, x|H_{T_{n-1}})1_{\{T_{n-1} < t \leq T_n\}} \quad (2.10)$$

and the cumulative hazard over the whole interval $[0, t]$ is

$$\Lambda_t(x) = \int_0^t d\Lambda_s(x). \quad (2.11)$$

2.3 The prediction process associated with a marked point process

The interpretation of the hazard as a probability of a new point in a "small interval dt"; i.e., $d\Lambda(t) = E(dN(t) = 1|\mathcal{F}_{t-}) = P(dN(t) = 1|\mathcal{F}_{t-})$, gives a very short term prediction. The questions raised in the first chapter concerned possible time lags between a cause and its effect. For this purpose we need long-term predictions of the effect. It turns out that a suitable notion is the prediction process associated with a marked point process. The properties of this process were studied in Norros (1985). We shall consider its usefulness in the analysis of causal chains. We follow in this section mainly the presentation in Arjas and Eerola (1993).

Consider the canonical space $(\Omega, \mathcal{F}_\infty)$ of the marked point process (T, X) and a random variable Y defined on this space, taking values in $(\mathbf{Y}, \mathcal{Y})$, where $\mathcal{Y}$ are the corresponding Borel sets. Define the *prediction process* $(\mu_t)_{t \geq 0}$ of Y by

$$\mu_t(I) = P(Y \in I | \mathcal{F}_t), \quad I \in \mathcal{Y}. \tag{2.12}$$

Letting now I vary in $\mathcal{Y}$ we obtain a distribution on $\mathcal{Y}$, and letting also t vary we obtain a distribution-valued stochastic process $(\mu_t)_{t \geq 0} = \{\mu_t(I); I \in \mathcal{Y}, t \geq 0\}$. It can be shown that there exist regular versions of transition probabilities

$$\mu_t^*(\cdot\,;\cdot) : \mathbb{H} \to \mathcal{Y}, \ t \geq 0,$$

such that we can use the representation

$$\mu_t(I) = \mu_t^*(H_t; I) \tag{2.13}$$

(i.e., $P(Y \in I | \mathcal{F}_t)(\omega) = \mu_t^*(H_t(\omega); I)$). We view Y here as the response, and therefore $\mu_t^*(H_t; \cdot)$ is the conditional distribution of the response given the pre-t history H_t.

In this framework a prediction, in terms of the process (μ_t^*), has always the following three parameters which can vary:

(1) given (t, H), the distribution $I \mapsto \mu_t^*(H; I)$ forms "in a given situation, a prediction concerning the value of Y";

(2) given (H, I), the function $t \mapsto \mu_t^*(H; I)$ shows how "the probability of $\{Y \in I\}$ is updated in time, based on the progressive observation of (H_t)" (*learning effect*);

(3) given (t, I), the function $H \mapsto \mu_t^*(H; I)$ shows how "the probability of $\{Y \in I\}$ depends on what has happened before t".

Each case (1)-(3) gives different information about the dynamics of the causal dependence, and this division turns out to be a convenient framework for the illustrations in Chapter 4.

We now show that the probability comparisons of Good (1961/62) and Suppes (1970), presented in Chapter 1, have their counterparts in the above dynamic framework. First identify the event $\{Y \in I\}$ with the response. Although in general Y can be $\mathbf{S} = (S_1, ..., S_k)'$ in the causal chain framework, so that $P(Y \in I | \mathcal{F}_t)$ gives a prediction of the "state of the chain at t, given the history up to t", we restrict usually Y to $S_k = S_E$, and then $I \subset R_+^1$. Let the strict pre-t history $H_{t-} = \{(T_n, X_n); T_n < t\}$ correspond to the background conditions B and the occurrence of a cause C correspond to a new point (t, x) appearing at t. Then the history at t becomes $H_{t-} \cup \{(t, x)\}$. On the other hand, if the cause C does not occur at t ($=\bar{C}$), the history remains H_{t-} also at t. The observation of (H_t) covers therefore both the relevant background information, the appearance of the cause, and ultimately also the response. The dynamic formulation of the difference $P(E|BC) - P(E|B\bar{C})$ in terms of the prediction process is then

$$\mu_t^*(H_{t-} \cup \{(t,x)\}; I) - \mu_t^*(H_{t-}; I) = C_t(x). \quad (2.14)$$

In order to find a similar dynamic reformulation to the difference $P(E|BC) - P(E|B)$ note first that the probability $P(E|B)$ can be written as $P(E|B) = P(C|B)P(E|BC) + P(\bar{C}|B)P(E|B\bar{C})$. It is therefore a weighted mean of $P(E|BC)$ and $P(E|B\bar{C})$, and it coincides with $P(E|BC)$ if $P(C|B) = 1$ and with $P(E|B\bar{C})$ if $P(\bar{C}|B) = 1$. In the latter case the two contrasts $P(E|BC) - P(E|B\bar{C})$ and $P(E|BC) - P(E|B)$ are the same.

Intuitively, the left limit $\mu_{t-}(I) = \mu_{t-}^*(H_{t-}; I)$ is seen to have the same role as $P(E|B)$ in Suppes' formulation. It corresponds to the situation when the occurrence of C is not yet known. We now show that $\mu_{t-}(I)$, too, has a weighted mean expression similar to that found for $P(E|B)$ above. For this purpose we study in more detail the behaviour of the process $\mu_t(I) = E(1_I(Y)|\mathcal{F}_t)$, $t \geq 0$, considering in particular its jumps $\Delta \mu_t(I) = \mu_t(I) - \mu_{t-}(I)$. Since $\mu_t(I) = \mu_t^*(H_t; I)$, it is clear that $\mu_t(I)$ equals $\mu_t^*(H_{t-} \cup \{(t,x)\}; I)$ if there is at t a marked point with mark x, and $\mu_t^*(H_{t-}; I)$ if there is no point at t. A coincise way of writing this is

$$\mu_t(I) = \sum_x \Delta N_t(x)\mu_t^*(H_{t-} \cup \{(t,x)\}; I) + (1 - \sum_x \Delta N_t(x))\mu_t^*(H_{t-}; I). \quad (2.15)$$

Similarly, considering the left limit $\mu_{t-}(I)$, one can show (see Arjas & Norros, 1984) that

$$\mu_{t-}(I) = \sum_x \Delta \Lambda_t(x)\mu_t^*(H_{t-} \cup \{(t,x)\}; I) + (1 - \sum_x \Delta \Lambda_t(x))\mu_t^*(H_{t-}; I). \quad (2.16)$$

where $\Delta \Lambda_t(x)$ has the interpretation $\Delta \Lambda_t(x) = P(\Delta N_t(x) = 1|\mathcal{F}_{t-})$. Therefore, $\mu_{t-}(I)$ is a weighted average of the possible predictions of $\{Y \in I\}$ at t, and the weights are the probabilities of the events $\{\Delta N_t(x) = 1\}$ and $\{\Delta N_t(x) = 0\}$. The above reasoning shows the exact correspondence between the difference $P(E|BC) - P(E|B)$ considered by Suppes (1970) and others, and the expression $\Delta \mu_t(I) = \mu_t(I) - \mu_{t-}(I)$ arising from our marked point process formulation, provided that $\Delta N_t(x) = 1$ and the connection between point (t,x) and cause C is made.

We now show the connection between the difference $C_t(x)$ in (2.14) and the progressive updating of the prediction of the event $\{Y \in I\}$ in (2) above. As an alternative to μ_t we may consider the conditional expectation of some real-valued and bounded test function h of Y. Then we have

$$M_t^h = E(h(Y)|\mathcal{F}_t) = \int_Y h(y)\mu_t^*(H_t; dy). \quad (2.17)$$

In our case always $h = 1_I$ so that $\mu_t(I) = P(Y \in I|\mathcal{F}_t) = E(1_I(Y)|\mathcal{F}_t) = M_t^{1_I}$. It can be shown (Arjas & Norros, 1984) that the process $(M_t)_{t\geq 0}$ (which in our case is the same as (μ_t)) is an $(\mathcal{F}_t)$-martingale and admits the *martingale representation*

$$\mu_t(I) = P(Y \in I) + \sum_x \int_0^t C_s(x)(dN_s(x) - d\Lambda_s(x)), \quad (2.18)$$

where $N_s(x) - \Lambda_s(x) = M_s(x)$ is the counting process martingale. The integrand $(C_t(\cdot))$ is called the *innovation gain* process, and $C_t(x)$ is the innovation gain from observing the point (t, x). This shows how the updating of the prediction in (2) above takes place. It is easy to see that if all compensators $\Lambda_t(x), x \in E$ are absolutely continuous (so that $\Delta\Lambda_t(x) = 0$) we have $\mu_{t-}(I) = \mu_t^*(H_{t-}; I)$, which is the same as the second term in (2.16). Then the points $(T_n, X_n)_{n\geq 1}$ are completely *unpredictable* from the $(\mathcal{F}_t)$-histories and process (μ_t) jumps only at the times $t = T_n$. The jumps are of the size

$$\Delta\mu_{T_n}(I) = \mu_{T_n}^*(H_{T_n-} \cup \{(T_n, X_n)\}; I) - \mu_{T_n-}^*(H_{T_n-}; I) = C_{T_n}(X_n). \quad (2.19a)$$

Between the jumps, i.e., when $T_{n-1} < t < T_n$, the prediction is updated according to

$$-C_t(x)d\Lambda_t(x). \quad (2.19b)$$

On the other hand, if the event $\{\Delta N_t(x) = 1\}$ can be predicted with certainty, so that $\Delta\Lambda_t(x) = 1$, we have $\Delta\mu_t(I) = 0$.

In the general case (when $\Delta\Lambda_t(x) \neq 0$) the innovation gain at t is as in (2.14). If $t = T_n$, the first term at right is $\mu_{T_n}^*(H_{T_n-} \cup \{(T_n, X_n)\}; I)$ and it coincides with the first term in (2.19a) whereas the second term in (2.14), $\mu_t^*(H_{t-}; I)$, equals with $\mu_{T_n-}^*(H_{T_n-}; I)$ in (2.19a) at $t = T_n$ only in the special case of continuous compensators since then $\mu_{T_n-}^*(H_{T_n-}; I) = \mu_{T_n}^*(H_{T_n-}; I)$.

The innovation gain $C_{T_n}(X_n)$ is the amount by which the prediction is corrected at $t = T_n$ when the marked point (T_n, X_n) is observed. In case (2) above, where we consider the function $t \mapsto \mu_t^*(H; I)$, the choice $t = T_n$ identifies $C_{T_n}(X_n)$ with the causal effect of (T_n, X_n) for the event $\{Y \in I\}$. Therefore, if the innovation gain $C_t(x)$ is positive for all t in the prediction interval, the occurrence of the cause increases the probability of the event $\{Y \in I\}$ for all t. Similarly, if $C_t(x)$ is negative for all t, it has preventive effect on $\{Y \in I\}$. Sometimes $C_t(x)$ can have both positive and negative values, or even the value 0 within a specified prediction interval. Then one can hardly say that the occurrence of (t, x) is a cause for $\{Y \in I\}$ in all circumstances. These are the cases in which the proposed method gives most new information compared to cross-sectional studies. Some applications with real data sets are considered in Chapter 4. We shall, however, first illustrate the ideas in a hypothetical example of a 3-point causal chain in which two causes act in a cumulating way. The example is from Arjas and Eerola (1993).

2.4 A hypothetical example of cumulating causes

Consider a causal chain: exposure $(A) \to$ disease $(B) \to$ death (C). Assume that the influences of exposure and disease on mortality are *linearly increasing in time*. This corresponds to the situation in which cumulative exposure of a (for example, carcinogenic) chemical raises the risk of a disease (e.g. cancer) in time. The disease, when becoming chronic, has then an additive excess risk on mortality. We simplify the setting by assuming that the hazards of exposure $\lambda_A(t)$ and disease $\lambda_B(t)$ are constant over time. Furthermore, we assume that $\lambda_{C|A}(t|v) = \lambda_C(t)$; i.e., exposure without subsequent illness has no effect on death hazard.

The event of prediction is survival at least until u; i.e., $\{T_C > u\}$, and it suffices to consider at time t histories of the form

$$\emptyset, \{(v, A)\}, \{(w, B)\}, \{(v, A), (w, B)\}$$

and "test sets" $I = (u, \infty], \ (0 < v < w < t < u)$.

The transition functions $\mu_t^*(H_t; I) = P(T_C > u | \mathcal{F}_t)$ have in the absolutely continuous case the explicit forms:

$$\mu_t^*(\{(v, A), (w, B)\}; (u, \infty]) = exp\{-\int_t^u \lambda_{C|A,B}(s|v, w)ds\}; \qquad (2.20)$$

$$\mu_t^*(\{(w, B)\}; (u, \infty]) = exp\{-\int_t^u \lambda_{C|B}(s|w)ds\}; \qquad (2.21)$$

$$\mu_t^*(\{(v, A)\}, (u, \infty]) = exp\{-\int_t^u [\lambda_{B|A}(s|v) + \lambda_{C|A}(s|v)]ds\}$$
$$+ \int_t^u \lambda_{B|A}(s|v)exp\{-\int_t^s [\lambda_{B|A}(r|v) + \lambda_{C|A}(r|v)]dr\} \qquad (2.22)$$
$$\times \mu_s^*(\{(v, A), (s, B)\}; (u, \infty])ds;$$

$$\mu_t^*(\emptyset; (u, \infty]) = exp\{-\int_t^u [\lambda_A(s) + \lambda_B(s) + \lambda_C(s)]ds\}$$
$$+ \int_t^u \lambda_A(s)exp\{-\int_t^s [\lambda_A(r) + \lambda_B(r) + \lambda_C(r)]dr\}\mu_s^*(\{(s, A)\}; (u, \infty])ds$$
$$+ \int_t^u \lambda_B(s)exp\{-\int_t^s [\lambda_A(r) + \lambda_B(r) + \lambda_C(r)]dr\}\mu_s^*(\{(s, B)\}; (u, \infty])ds.$$
$$(2.23)$$

Table 1. The functional forms of the hazards

$$\lambda_A(t) = \frac{1}{100};$$
$$\lambda_B(t) = \frac{1}{200};$$
$$\lambda_C(t) = \frac{1}{200} + \frac{1}{200}t;$$
$$\lambda_{B|A}(t|v) = \lambda_C(t) + \frac{1}{40} + \frac{1}{400}(t-v);$$
$$\lambda_{C|A}(t|v) = \lambda_C(t);$$
$$\lambda_{C|B}(t|w) = \lambda_C(t) + \frac{1}{25} + \frac{1}{250}(t-w);$$
$$\lambda_{C|A,B}(t|v,w) = \lambda_{C|B}(t|w).$$

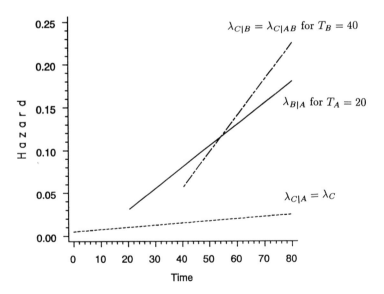

Fig. 2.2. Conditional hazards

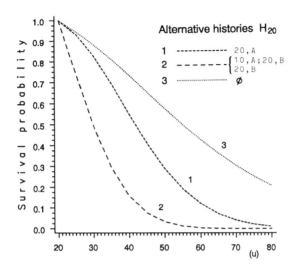

Fig. 2.3. Survival probabilities until age u, evaluated at $t = 20$

We follow here the division (1)-(3) in Section 2.3 and let the starting point t, history H and endpoint u of the prediction vary one at a time, and keep the other two fixed. Consider first the situation (case 1) in which the *starting point of the prediction ("the present") has a fixed value at* $t = 20$ *and the endpoint u varies in the interval* $u \in (t, 80]$; i.e., the mapping $u \mapsto \mu_t^*(H; (u, \infty])$, given (t, H). This is the prediction concerning the event $\{T_C > u\}$, given the prespecified history (Fig. 2.3).

At each u-value one has the point prediction of survival at least until u, given the observed history at $t = 20$. Different 20-year long histories give rise to different survival curves. The assumption of an increasing effect of A and B on C in time results in a rapidly decreasing survival curve after $t = 20$ if A or B has occurred previously. The survival probability to age 60 is practically zero for those who were diagnosed at age 20, whereas there is a 20% chance that an individual who was neither exposed nor diseased at that age is still alive at age 80. Note that the curves for $\mu_t^*(\{(v, A), (w, B)\}; (u, \infty])$ and $\mu_t^*(\{(w, B)\}; (u, \infty])$ are identical because $\lambda_{C|A,B}(t|v, w)$ was assumed to be the same as $\lambda_{C|B}(t|w)$.

Obviously only some specific histories can be considered in this way, but the question "Do different occurrence times of A and B lead to different conclusion?" cannot be dealt with satisfactorily. It appears that an explicit consideration of the innovation gain process, corresponding to a particular mark, offers a useful way for such an analysis. Consider, therefore, next what happens when *the time at which the prediction is made varies* in the interval $[0, 80]$; i.e., given (H, I), we consider $t \mapsto \mu_t^*(H; I)$.

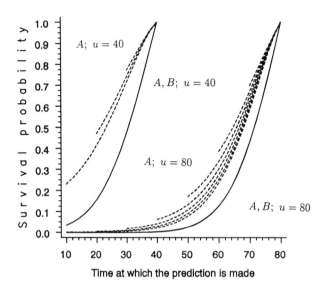

Fig. 2.4. Survival probabilities until age u, (1) when A has occurred anytime previously and B has just occurred (solid line), (2) when A has occurred previously at times $T_A = 0, 10, ..., u - 10$ and B has not yet occurred (broken lines), each corresponding to a particular value of T_A.

In case (2) we compare the cross-sections of the prediction curves, or in fact the point predictions at a specified u-value (here at $u = 40, 80$) for all t-values before u. We also consider the predictions of the form $t \mapsto \mu_t^*(H \cup \{(t, B)\}; I)$, where H is either $\{\emptyset\}$ or $\{(v, A)\}$ with $v < t$ corresponding to situations where B has just occurred. The differences $\mu_t^*(H \cup \{(t, B)\}; (u, \infty]) - \mu_t^*(H; (u, \infty])$ represent then the innovation gains from observing B at t, when predicting the event $\{T_C > u\}$. We have two sources of variation here. First, the longer the prediction interval, the longer lasts the cumulating effects of A and B. Second, the longer the prediction interval, the higher is the probability of dying in that interval. We consider here only the most complicated case in which either both A or B have already occurred, or A has occurred and the occurrence of B is still possible.

When the effect of the causes A and B on C increases in time and the prediction concerns survival to 80 years, the probability of surviving remains close to zero until $T_B = t$ is almost 45 years (Fig. 2.4). After that it rapidly increases and eventually equals 1. This final value expresses the trivial fact that if an exposed individual remains alive and disease-free until the age of 80, the survival to at least 80 years is certain, no matter whether $T_B = 80$ or not.

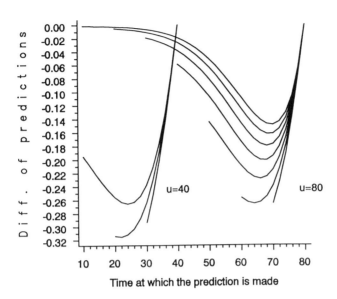

Fig. 2.5. Innovation gains in predicting survival until age u, from observing B when A has occurred previously. Each curve corresponds to a particular value of T_A ($T_A = 0, 10, ..., u - 10$).

The innovation gains from observing B, shown in Figure 2.5, demonstrate the counterfactual interpretation: "given that A has occurred previously, how do the conditional survival probabilities to u differ in the two situations where (i) also B has just occurred, and (ii) B did not occur?"

The differences of observing (resp. not observing) B at t, given that A occurred at v are

$$\mu_t^*(\{(v,A),(t,B)\};(u,\infty]) - \mu_t^*(\{(v,A)\};(u,\infty]); \quad v \le t \le u.$$

Since the event of interest is here the nonoccurrence of C, the differences are negative if B is a risk factor of C, and positive if it has a preventive effect.

Consider finally case (3) in which *the prediction interval from t to u is fixed but the history before t is allowed to vary*. In this way we can eliminate the source of variation which comes from the varying length of the prediction interval (Fig. 2.6). Given (t, I) with $I = (u, \infty]$, we consider the mapping $H \mapsto \mu_t^*(H; I)$. We fix here $t = 35$ and $u = 40, 60$ or 80, and let T_A and T_B vary in the interval $[0, 35]$. Compared to the previous predictions, we now consider the net influence of the occurrence times of A and B because the length of the prediction interval is held fixed.

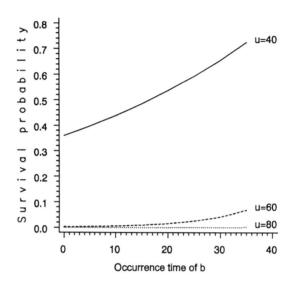

Fig. 2.6. Dependence of the survival probability until age u on the occurrence time of B when A has occurred any time previously. Time of prediction=35.

The time of diagnosis has a very strong influence on survival when the prediction interval is short (from 35 to 40 years). Early disease onset decreases the survival probability drastically compared to onset occurring only a few years before the time of the prediction at $t = 35$ years. This is, of course, expected because the harmful effects from the disease have had little time to cumulate. When the prediction interval is maximal (from 35 to 80 years), the occurrence time of B seems completely unimportant. Practically all diseased individuals are expected to die before age of 80 regardless of the time of onset. In the intermediate case (prediction from 35 to 60 years) some 5 per cent of the individuals with onset at 35 are expected to stay alive to age 60.

2.5 Causal transmission in terms of the prediction process

We reconsider briefly the concept of causal transmission in a causal chain. As is obvious by now, the conditional probabilities of the response E can be expressed dynamically in terms of the prediction process $\mu_t(I)$ of S_E. The influence of a particular event C_r, when the occurrence of all other causes is arbitrary, can be studied by comparing those predictions of E in which C_r either has just occurred or has not yet occurred; i.e., by considering the innovation gains $C_t(x)$, where $t = S_r$ and $x = C_r$.

Consider the special case of *causative* (positive) effects in a general k-point causal chain, and in particular, causal transmission from C_1 to $C_k = E$. In terms of probabilities, we obviously have by the theorem of total probability

$$P(E|C_1) = P(C_2|C_1)P(C_3|C_1C_2)...P(C_{k-1}|C_1C_2...C_{k-2})P(E|C_1C_2...C_{k-1})$$
$$+ P(C_2|C_1)...(1 - P(C_{k-1}|C_1C_2...C_{k-2}))P(E|C_1C_2...\bar{C}_{k-1})$$
$$\vdots$$
$$+ (1 - P(C_2|C_1))...(1 - P(C_{k-1}|C_1\bar{C}_2...\bar{C}_{k-2}))P(E|C_1\bar{C}_2..\bar{C}_{k-1}).$$
(2.24)

Similarly,

$$P(E|\bar{C}_1) = P(C_2|\bar{C}_1)P(C_3|\bar{C}_1C_2)...P(C_{k-1}|\bar{C}_1C_2...C_{k-2})P(E|\bar{C}_1C_2...C_{k-1})$$
$$+ P(C_2|\bar{C}_1)...(1 - P(C_{k-1}|\bar{C}_1C_2...C_{k-2}))P(E|\bar{C}_1C_2...\bar{C}_{k-1})$$
$$\vdots$$
$$+ (1 - P(C_2|\bar{C}_1))...(1 - P(C_{k-1}|\bar{C}_1\bar{C}_2...\bar{C}_{k-2}))P(E|\bar{C}_1\bar{C}_2...\bar{C}_{k-1}).$$
(2.25)

As in Section 2.3, the corresponding dynamic form of the probability $P(E|C_1)$ in terms of the prediction process of S_E is now $\mu_t^E(\{(t, C_1)\}; I)$, and that of $P(E|\bar{C}_1)$ is $\mu_t^E(\emptyset; I)$. Then the condition $P(E|C_1) > P(E|\bar{C}_1)$ is expressed in terms of the prediction process in the following way: If, for all t,

$$\mu_t^E(\{(t, C_1)\}; I) > \mu_t^E(\emptyset; I) \qquad (2.26)$$

then C_1 has causative effect on the occurrence of E. The prediction $\mu_t^E(\{(t, C_1)\}; I)$ is a sum over all possible paths in which C_1 has occurred at t, and $\mu_t^E(\emptyset; I)$ a sum over those paths in which C_1 has not occurred. More generally, we may of course consider an arbitrary event $C_r, r = 1, ..., k - 1$ in (2.26) instead of C_1. Then the corresponding condition is (as in Section 2.3)

$$\mu_t^E(H_{t-} \cup \{(t, C_r)\}; I) > \mu_t^E(H_{t-}; I), \quad \text{for all } t, r = 1, ..., k-1, \qquad (2.27)$$

with the convention that when $r = 1$, $H_t = \emptyset$, as in (2.26). The requirement that (2.27) holds for all r corresponds to the notion of a system *weakened by failures* (Arjas & Norros, 1984, Norros, 1985). Then, at each occurrence time S_i, the probability that the k-component system still functions decreases; i.e., $\mu_{S_i} \leq \mu_{S_i-}$, for all i. Nothing else is assumed about the mutual dependence of the components. Neither do we assume anything about the dependence between C_r and $C_j, r, j = 1, ..., k-1, r \neq j$ in (2.27). In general, the condition (2.27) may not hold for all t, indicating that the influence of a cause is transient (when letting I vary in $\mathcal{Y}$), or that it even turns out to be preventive at some t.

3. CONFIDENCE STATEMENTS ABOUT THE PREDICTION PROCESS

3.1 Introduction

So far our approach has been purely probabilistic. The hypothetical hazards in the example in Chapter 2 were chosen appropriately for illustration purposes. However, in order to apply the method in statistical analysis with real data sets, we must specify how the hazards are estimated from the data. This amounts to specifying a statistical model for the hazards. Hazard regression models are standard tools in modelling dependence between hazard related to a response variable and a set of explanatory variables. As an example, we consider in this work the *discrete time logistic regression model*. In general, the way the hazards are estimated is a secondary question in a further analysis of the prediction probabilities, but it is of course a crucial part of relevant data analysis. We return to that in Chapter 4.

The main reason for using the logistic regression model as an example is that in a fully parametric model well known asymptotic results can be used in a straightforward (although elaborate) manner when considering the statistical properties of the prediction probabilities. The structure of the logistic model also corresponds to the natural idea of comparing the probabilities of success and failure in a series of successive Bernoulli trials. Continuous time models can be approximated as closely as one wants to by choosing a small grid interval for the time variable. The underlying follow-up time, which has to be parameterized also, can be modelled in an arbitrary way, in the simplest case as indicator variables which are constant within an interval.

In this chapter we shall consider different ways of constructing confidence intervals for the prediction probabilities. At least four different approaches can be mentioned: First, the asymptotic variance of the prediction probabilities can be derived by the generalized delta-method. It is well known that under suitable regularity conditions the asymptotic normality of a statistic implies asymptotic normality also for its functionals. The second method is to use the upper and lower confidence limit values for the hazard model parameters as plug-in values in the prediction formulae. In order to get confidence sets which are intervals, we need, however, to check that the prediction probabilities are monotonic functions of the hazards. If this holds, results concerning the stochastic order of random life length vectors can be applied. The third approach, which has become extensively used nowadays, is to apply resampling methods like bootstrapping. Since the hazard models include stochastic covariates, the generation of all possible samples seems, however, overwhelmingly elaborate in our case. The fourth possibility is to use Bayesian methods to obtain a posterior distribution of the parameters and thus that of the prediction probabilities. We shall restrict ourselves here to the first two approaches.

3.2 Prediction probabilities in the logistic regression model

We start by showing how the prediction probabilities are expressed in terms of the hazards in the logistic regression model. In the causal chain framework of Chapter 2 it is assumed that all randomness is in the occurrence times of the causal events $C_1, C_2, ..., C_k$. Even though other explanatory variables in the hazard models can be time-varying, they are assumed fixed at the time of conditioning (i.e., they are $\mathcal{F}_0$-measurable), following the convention of ordinary regression analysis. We consider here the simplest 3-point causal chain but the results apply to a general k-point chain as well. The events are denoted by A, B and C, their occurrence times by T_A, T_B and T_C, and the corresponding model parameter vectors by α, β and γ. We consider the logistic regression model in terms of counting processes N_A, N_B and N_C which count the occurrences of the causal events (e.g. $N_t(A) = \sum_{n \geq 1} 1_{\{T_n \leq t, X_n = A\}}$). The logistic model for the event B, for example, is then of the form

$$ln(\frac{P(\Delta N_t(B) = 1|\mathcal{F}_{t-1})}{P(\Delta N_t(B) = 0|\mathcal{F}_{t-1})}) = \beta' Z_{t-1}(B), \quad (3.1)$$

if B has not yet occurred, and the B-specific hazard is

$$p_B(t) = P(\Delta N_t(B) = 1|\mathcal{F}_{t-1}) = Y_{t-1}(B) \frac{e^{\beta' Z_{t-1}(B)}}{1 + e^{\beta' Z_{t-1}(B)}}, \quad (3.2)$$

where the indicator $Y_t(B) = 1$ if B and the subsequent events in the causal chain (in this case C) have not occurred by t (or at t), and 0 otherwise.

The failure pattern specific hazards in the 3-point hierarchical model can furthermore be written in terms of the occurrence times $T_A \leq T_B \leq T_C$ as follows

$$\begin{aligned}
p_A(t) &= P(T_A = t | T_A \geq t, T_B \geq t, T_C \geq t, .), \\
p_B(t) &= P(T_B = t | T_A \geq t, T_B \geq t, T_C \geq t, .), \\
p_C(t) &= P(T_C = t | T_A \geq t, T_B \geq t, T_C \geq t, .). \\
p_{B|A}(t|v) &= P(T_B = t | T_A = v, T_B \geq t, T_C \geq t, .), \quad (3.3) \\
p_{C|A}(t|v) &= P(T_C = t | T_A = v, T_B \geq t, T_C \geq t, .), \\
p_{C|B}(t|w) &= P(T_C = t | T_A = \infty, T_B = w, T_C \geq t, .), \\
p_{C|AB}(t|v, w) &= P(T_C = t | T_A = v, T_B = w, T_C \geq t, .).
\end{aligned}$$

The dot "." means the additional, at time t fixed, conditioning variables in a particular hazard model. The hazards conditioned on A and/or B include time-dependent covariates which measure the effect of the occurrence of these preceding causal events; for example, in $p_{B|A}(t|v)$ the value of the covariate $Z_t(A) = 1, t \geq v$, when A has occurred at v, and zero before that, whereas in $p_B(t)$ the covariate $Z_t(A)$ is not defined.

For convenience, we consider, as in Chapter 2, the nonoccurrence of the response C. Thus, in discrete time the prediction $P(T_C > u|\mathcal{F}_t) = \mu_t^*(H_t; (u, \infty])$, given the four possible recorded histories before t, has the following form (cf. Chapter 2):

For $0 < v < w < t < u$,

$$P(T_C > u|T_A = v, T_B = w, T_C > t) = \prod_{s=t+1}^{u} (1 - p_{C|AB}(s|v,w)) \tag{3.4}$$

$$P(T_C > u|T_A = \infty, T_B = w, T_C > t) = \prod_{s=t+1}^{u} (1 - p_{C|B}(s|w)); \tag{3.5}$$

$$\begin{aligned}P(T_C > u|T_A = v, T_B > t, T_C > t) &= \prod_{s=t+1}^{u} (1 - p_{B|A}(s|v) - p_{C|A}(s|v)) \\ &+ \sum_{s=t+1}^{u} \prod_{r=t+1}^{s-1} (1 - p_{B|A}(r|v) - p_{C|A}(r|v)) p_{B|A}(s|v) \\ &\quad \times P(T_C > u|T_A = v, T_B = s, T_C > s);\end{aligned} \tag{3.6}$$

$$\begin{aligned}P(T_C > u|T_A > t, T_B > t, T_C > t) &= \prod_{s=t+1}^{u} (1 - p_A(s) - p_B(s) - p_C(s)) \\ &+ \sum_{s=t+1}^{u} \prod_{r=t+1}^{s-1} (1 - p_A(r) - p_B(r) - p_C(r)) p_A(s) \\ &\quad \times P(T_C > u|T_A = v, T_B > s, T_C > s) \\ &+ \sum_{s=t+1}^{u} \prod_{r=t+1}^{s-1} (1 - p_A(r) - p_B(r) - p_C(r)) p_B(s) \\ &\quad \times P(T_C > u|T_A > s, T_B = s, T_C > s),\end{aligned} \tag{3.7}$$

in which the hazards $p.(t)$ are of the form (3.2).

3.3 Confidence limits for μ_t using the delta-method

Arjas and Haara (1986) (see also Hauck, 1983) showed that under suitable regularity conditions the ML-estimators $\hat{\beta}$ in the logistic regression model are consistent and asymptotically normal. They in fact showed that $\sqrt{c_t}(\hat{\beta} - \beta) \xrightarrow{D} N(0, \Sigma)$, as $t \to \infty$, where c_t can be the total time on test at t, for example, but the result holds also in the more usual formulation when the risk set $n \to \infty$. It is well known that the asymptotic distribution of a function g of an asymptotically normal random variable $\hat{\theta}$, say, can be obtained by the delta method from the Taylor expansion of g around θ, provided that g is at least twice differentiable at θ. Since $\sqrt{n}(g(\hat{\theta}) - g(\theta)) = \sqrt{n}(\hat{\theta} - \theta)g'(\theta) + o_p(1)$, it follows that

$$\sqrt{n}(g(\hat{\theta}) - g(\theta)) \xrightarrow{D} N(0, \sigma^2(g'(\theta))^2). \tag{3.8}$$

where σ^2 is the asymptotic variance of $\hat{\theta}$. More generally, if $\mathbf{g}(\hat{\theta}) = (g_1(\hat{\theta}), g_2(\hat{\theta}), ..., g_k(\hat{\theta}))'$, where $\hat{\theta}$ is a p-vector and $(\partial g/\partial \theta)$ is a $(k \times p)$-matrix with elements $\partial g_i(\theta)/\partial \theta_j$ at $\hat{\theta} = \theta$, we have

$$\sqrt{n}(\mathbf{g}(\hat{\theta}) - \mathbf{g}(\theta)) \xrightarrow{D} N(0, (\partial g/\partial \theta)\Sigma(\partial g/\partial \theta)'), \tag{3.9}$$

where Σ is the asymptotic covariance matrix of $\hat{\theta}$. In our case, $\mathbf{g}(\theta)$ are complicated functions of the model parameters $\theta = (\alpha, \beta, \gamma)$, as is obvious from (3.4)-(3.7), and thus the derivatives of $\mathbf{g}(\theta)$ in the 3-point dynamic model are also quite involved. Otherwise, since we are dealing with a fully parametric model, the delta-method and the result in (3.9) can be applied in a straightforward manner.

Fix now the endpoint u. Define $F_s(\theta) = P(T_C > s | \mathcal{F}_t), t < s \leq u$, and $\theta = (\alpha, \beta, \gamma)$. Then, for any $r \leq s$, such that $t + 1 \leq r \leq s \leq u$, the estimated asymptotic covariance of $F_r(\theta)$ and $F_s(\theta)$ is

$$\hat{cov}(F_r(\hat{\theta}), F_s(\hat{\theta})) = \sum_{k=t+1}^{s} \sum_{l=t+1}^{r} F_k'(\hat{\theta}) \hat{cov}(\hat{\theta}) (F_l'(\hat{\theta}))', \tag{3.10}$$

where $F_k'(\hat{\theta}) = \left(\partial F_k(\hat{\theta})/\partial \theta_1, ..., \partial F_k(\hat{\theta})/\partial \theta_p\right)'$. For each probability in (3.4)-(3.7), the formulas of the derivatives $\partial F_k(\hat{\theta})/\partial \theta_j$ are given in Appendix 1. In the hierarchic 3-point chain model the asymptotic covariance matrix of the parameter vector $\hat{\theta} = (\hat{\alpha}, \hat{\beta}, \hat{\gamma})$ is a block-diagonal matrix since the parameters are estimated independently; i.e.,

$$\hat{cov}(\hat{\theta}) = I^{-1}(\hat{\theta}) = \begin{pmatrix} I^{-1}(\hat{\alpha}) & 0 & 0 \\ 0 & I^{-1}(\hat{\beta}) & 0 \\ 0 & 0 & I^{-1}(\hat{\gamma}) \end{pmatrix} \tag{3.11}$$

where

$$I(\hat{\theta}) = \left\{ -\frac{\partial^2}{\partial \theta_k \theta_l} \log L(\theta) \right\}_{|\theta = \hat{\theta}}.$$

Each component of $F'(\hat{\theta})$, corresponding to $\hat{\alpha}, \hat{\beta}$ and $\hat{\gamma}$, can therefore be calculated and multiplied separately by the corresponding information matrix. Sometimes it may be more natural to introduce parameters which are common to all hazard models $(A, B$ and $C)$. In these cases the off-diagonal terms of the covariance matrix are of course not zero.

Pointwise asymptotic $(1-p)$-level confidence intervals for $F_s(\theta) = P(T_C > s|\mathcal{F}_t), t < s \leq u$, are then obtained from

$$F_s(\hat{\theta}) \pm c_{p/2}\sqrt{v\hat{a}r(F_s(\hat{\theta}))/n}, \tag{3.12}$$

where $c_{p/2}$ is the $p/2$-quantile of the standard normal distribution and $v\hat{a}r(F_s(\hat{\theta}))$ is the sum of the diagonal elements of (3.10) at $v = t+1, ..., s$; i.e.,

$$v\hat{a}r(F_s(\hat{\theta})) = \sum_{v=t+1}^{s} \sum_{j} (\frac{\partial F_v(\hat{\theta})}{\partial \theta_j})^2 v\hat{a}r(\hat{\theta}_j) + \sum_{v=t+1}^{s} \sum_{\substack{j,k \\ j \neq k}} (\frac{\partial F_v(\hat{\theta})}{\partial \theta_j}) c\hat{o}v(\hat{\theta}_j, \hat{\theta}_k)(\frac{\partial F_v(\hat{\theta})}{\partial \theta_k}).$$
$$\tag{3.13}$$

Since the component vectors of $\theta = (\alpha, \beta, \gamma)$ are estimated independently, each quadratic form can actually be calculated and added separately. For example, in (3.2) and (3.3) we simply have $\theta = \gamma$.

Denote $F_u(\hat{\theta}(x)) = \hat{\mu}_t^*(H_{t-} \cup \{(t,x)\}; (u, \infty])$ and $F_u(\hat{\theta}(\emptyset)) = \hat{\mu}_t^*(H_{t-}; (u, \infty]), t \leq u$. The asymptotic variance of the estimated innovation gain

$$\hat{C}_t(x) = \hat{\mu}_t^*(H_{t-} \cup \{(t,x)\}; (u, \infty]) - \hat{\mu}_t^*(H_{t-}; (u, \infty])$$

from observing event $x = C_r$ at t is then estimated by

$$v\hat{a}r(\hat{C}_t(x)) = \frac{n_1}{n_1 + n_2} v\hat{a}r(F_u(\hat{\theta}(x))) + \frac{n_2}{n_1 + n_2} v\hat{a}r(F_u(\hat{\theta}(\emptyset))) \tag{3.14}$$

where n_1 is the sample size of those who experienced event $x = C_r$ and n_2 the sample size of those who did not. The $(1-p)$-level confidence limits for $C_t(x)$ can be obtained from

$$\hat{C}_t(x) \pm c_{p/2}\sqrt{v\hat{a}r(\hat{C}_t(x))/K}, \tag{3.15}$$

where $K = n_1 n_2/(n_1 + n_2)$.

The above results provide *pointwise* confidence intervals for the prediction probabilities. In longitudinal studies more interesting regions are, however, often those which include, for example, the survival distribution for all t within a certain time interval. In our case the natural interval is the prediction interval $(t, u]$. This is obviously a more difficult result to obtain. There is a considerable amount of literature concerning *confidence bands* for the survival distribution in the non-parametric case (e.g. Breslow & Crowley, 1974, Gill, 1980, Hall & Wellner, 1980, Nair, 1984). These works are based on results concerning the weak convergence of processes of the form $g_n(\hat{F} - F)$, where g_n is a nonnegative function and $\hat{F}$ is the empirical distribution function. When $g_n(\hat{F} - F)$ converges on some interval

$[0, t]$ to a limit process Q, then by the Continuous Mapping Theorem also for the supremum holds

$$\sup_{0 \leq s \leq t} g_n(s)|\hat{F}(s) - F(s)| \xrightarrow{D} \sup_{0 \leq s \leq t} |Q(s)|. \qquad (3.16)$$

If $q_p(t)$ now satisfies

$$P\Big(\sup_{0 \leq s \leq t} |Q(s)| \leq q_p(t)\Big) \geq 1 - p \qquad (3.17)$$

then asymptotically

$$P\Big(\sup_{0 \leq s \leq t} g_n(s)|\hat{F}(s) - F(s)| \leq q_p(t)\Big) \geq 1 - p, \qquad (3.18)$$

so that

$$P\Big(\hat{F}(s) - \frac{q_p(t)}{g_n(s)} \leq F(s) \leq \hat{F}(s) + \frac{q_p(t)}{g_n(s)}; 0 \leq s \leq t\Big) \geq 1 - p \qquad (3.19)$$

and thus $\hat{F}(s) \pm [q_p(s)(g_n(s))^{-1}]$, $0 \leq s \leq t$ is an approximate $(1-p)$-level confidence band for F on $[0,t]$ (e.g. Fleming & Harrington, 1990).

In the censored case a nonparametric estimator of the survival distribution $S = 1 - F$ is the Kaplan-Meier estimator

$$\hat{S}(t) = \prod_{s \leq t} \Big(1 - \frac{\Delta N_s}{Y_s}\Big), \qquad (3.20)$$

where the process Y_t is the size of the risk set at t. When deriving the large sample distribution of $\hat{F}$ a result of Gill (1980) concerning the asymptotics of certain k-sample statistics plays an important role. Gill provided conditions for the weak convergence of processes of type $U_t(i) = \int_0^t H_s(i) dM_s(i), i = 1, ..., k$, where $H(i)$ are locally bounded predictable processes and $M(i) = N(i) - A(i)$ are counting process martingales. Since $\sqrt{n}(S(t) - \hat{S}(t))$ can be written in the form

$$\sqrt{n}(\hat{S}(t) - S(t)) = -S(t) \int_0^t \frac{\hat{S}(s-)}{S(s)} \sqrt{n} \left[\frac{dN_s}{Y_s} - 1_{\{Y_s > 0\}} d\Lambda_s - 1_{\{Y_s = 0\}} d\Lambda_s\right], \qquad (3.21)$$

where dN_s/Y_s is the Nelson-Aalen estimator of the cumulative hazard Λ_s, and the process

$$H_s = \sqrt{n} \frac{\hat{S}(s)}{S(s)} \frac{1_{\{Y_s > 0\}}}{Y_s} \qquad (3.22)$$

is predictable and locally bounded, the result of Gill (1980) can be applied. The bias term $\sqrt{n} B_t = -S(t) \int_0^t \hat{S}(s-)/S(s) 1_{\{Y_s = 0\}} d\Lambda_s$ turns out to be asymptotically negligible, and it can be shown (e.g. Fleming & Harrington, 1990, Theorem 6.3.1) that in a random censorship model

$$\sqrt{n}(\hat{F}(\cdot) - F(\cdot)) \Longrightarrow (1 - F(\cdot))W(v(\cdot)) \quad \text{on } D[0,t], \qquad (3.23)$$

where " $\Longrightarrow$ " denotes weak convergence, W is the Brownian motion and $D[0,t]$ the space of right continuous functions with left limits. The term $v(t) = \int_0^t (\pi(s))^{-1} d\Lambda_s$ is the asymptotic variance of $\hat{S}(t)$ where the probability of "staying at risk" is $\pi(s) = P(T^* > 0)$, with $T^* = min(T, C)$ and C the censoring time.

A more useful result as regards to confidence bands is, however, the following part of the same theorem

$$\sup_{0 \leq s \leq t} \left\{ \frac{n}{\hat{v}(t)} \right\}^{1/2} \frac{|\hat{F}(s) - F(s)|}{1 - \hat{F}(s)} \xrightarrow{\mathcal{D}} \sup_{0 \leq x \leq 1} |W(x)|, \qquad (3.24)$$

where $\hat{v}(t) = n \int_0^t [(Y_s - \Delta N_s) Y_s]^{-1} dN_s$ is the estimator of $v(t)$ (so that $\hat{S}^2(t) \hat{v}(t)$ is the Greenwood's formula). This can be recognized as a weighted version of the Kolmogorov-Smirnov test statistic.

The result in (3.24) is useful when constructing confidence bands for F since the distribution of $\sup_{0 \leq x \leq 1} |W(x)|$ is known and its values are tabulated. The $(1-p)$-confidence band for $F(t)$ can therefore be obtained from

$$\hat{F}(s) \pm \phi_{1-p} (1 - \hat{F}(s)) \left\{ \frac{\hat{v}(t)}{n} \right\}^{1/2}, \ 0 \leq s \leq t, \qquad (3.25)$$

where ϕ_{1-p} is the $(1-p)$-quantile of that distribution. These confidence bands can be constructed only for a time interval $[0, t]$ where $P\{\pi(s) > 0\}$, $0 \leq s \leq t$. This means in practice that $t = T_n$. It is not a restriction in our case since the prediction interval extends only to the last observed occurrence time in the data set.

The derivation of the confidence bands in the non-parametric case relies on the fact that the difference $\sqrt{n}(\hat{F}(t) - F(t))$ can be expressed in the form of a martingale $U_t = \int H_s M_s$ in which the weighting process H_t is locally bounded and predictable. This is a consequence of the structure of the Nelson-Aalen estimator of the cumulative hazard Λ_t. In our case the hierarchy in the causal chain is modelled through the hazards and the history is expressed in terms of the stochastic covariates Z_t. This means that our approach must be at least partly parametric. In non-parametric analysis we would need a separate counting process for each event, given an arbitrary history of possible occurrence times of T_A and T_B. The asymptotic variance of the non-parametric Kaplan-Meier estimator which appears in (3.25) serves, however, as an upper limit to the asymptotic variance of the binomial logistic model with $p = n$, and with the choice $\mathbf{Z}_{p \times p} = \mathbf{I}_{p \times p}$ of the design matrix, (Efron, 1988); i.e., when there is a parameter for each interval, the number of which is the same as the number of failures used in the Kaplan-Meier estimator. It is therefore expected that the confidence bands of the logistic model lie between the bands obtained from the non-parametric analysis, and when $p \to n$, the asymptotic variance of $\hat{S}(t)$ of the logistic model approaches the asymptotic variance of the Kaplan-Meier estimator.

Andersen *et al.* (1991) considered a similar three-state model and used the semiparametric Cox regression model for the hazards in which the previous history was also modelled in terms of covariates. They derived the large sample properties of the survival probabilities from the properties of $(\hat{\beta}, \hat{\Lambda}_t(\hat{\beta}))$, which were given in Andersen and Gill (1982). The non-parametric form of $\hat{\Lambda}_t(\hat{\beta})$ provides in this case the link to martingale techniques. The main difference, as regards our more general model, is, however, that they assumed a simplifying Markov-property by cutting the conditioning history. This assumption is unrealistic in complicated causal chains.

A practical way to derive (conservative) confidence bands for the prediction probabilities or for the innovation gains is to use the Bonferroni inequality. The critical value $c_{p/2}$ is then increased to $c_{p/2k}$, where k is the number of equally spaced time points for which the intervals are simultaneously valid (e.g. Dabrowska *et al.*, 1992).

We consider next a more straightforward way of constructing confidence limits for the prediction probabilities which is possible if the hazards in the causal chain are *monotonic* functions of the histories. However, since the prediction probabilities are complicated functions of the regression parameters there is no guarantee that the resulting confidence sets are *intervals* anymore. In the next section we make the additional assumption of monotonic hazards. This turns out to be a sufficient condition for monotonic prediction probabilities.

3.4 Confidence limits for μ_t based on the monotonicity of hazards

In this second approach we derive a simultaneous confidence band for the linear combination $\beta'\mathbf{z}$ and use the limits directly as plug-in values in the hazards, and consequently, in the prediction formulae. When the confidence limits for the hazards are given, an order relation between two hazard functions turns out to offer an easily verifiable way to check the monotonicity of the corresponding prediction probabilities as functions of the history. We emphasize, however, that these confidence limits require the assumption of monotonic hazards which is valid in hierarchic models where there is a positive effect between each event. This assumption is not always realistic.

We begin with a classical result of simultaneous inference in Gaussian regression models. Consider the model

$$\mathbf{y} = \mathbf{Z}\beta + \epsilon, \tag{3.26}$$

where $\mathbf{Z}$ is the $(n \times p)$-design matrix with $\text{rank}(\mathbf{Z}) = p$, $\beta = (\beta_1, ..., \beta_p)$ is a vector of fixed unknown parameters and the components of $\mathbf{z} = (z_1, ..., z_n)$ are known real values. The error terms are assumed to satisfy $\epsilon \sim N(0, \sigma^2)$ where the unknown variance σ^2 is estimated by $s^2 = \mathbf{y}(\mathbf{I} - \mathbf{Z}'(\mathbf{Z}'\mathbf{Z})^{-1}\mathbf{Z})/(n-p)$. The estimators $\hat{\beta}$ and s^2 are distributed as follows:

$$\hat{\beta} \sim N(\beta, \sigma^2(\mathbf{Z}'\mathbf{Z})^{-1}) \quad \text{and} \quad \frac{(n-p)s^2}{\sigma^2} \sim \chi^2_{n-p}, \tag{3.27}$$

with $\hat{\beta}$ and s^2 independent.

Then it is well-known (Working-Hotelling, 1929 and Scheffé, 1953) that the interval

$$\hat{\beta}'\mathbf{z} \pm c_{\alpha,p} s (\mathbf{z}'(\mathbf{Z}'\mathbf{Z})^{-1}\mathbf{z})^{1/2} \qquad (3.28)$$

provides a $(1-\alpha)$-*confidence band* for the linear combination $\beta'\mathbf{z}$ for all $\mathbf{z} \in R^p$. The constant c is chosen so that

$$P(|\beta'\mathbf{z} - \hat{\beta}'\mathbf{z}| \leq c_{\alpha,p} s^2 \mathbf{z}'(\mathbf{Z}'\mathbf{Z})^{-1}\mathbf{z}, \quad \forall \mathbf{z} \in R^p) = 1 - \alpha. \qquad (3.29)$$

From the distributional assumptions in (3.27) it follows that the constant is given by $c_{\alpha,p} = p F_{\alpha,p,n-p}$, where F the critical value of the F-distribution. The interval (3.29) is the projection of the confidence ellipsoid, centered at $\hat{\beta}$, onto the 1-dimensional subspace which is generated by $\mathbf{z}$. The tangent planes of the ellipsoid, which are orthogonal to this subspace, restrict the length of the confidence interval of $\beta'\mathbf{z}$.

These so called *Scheffé-type confidence bands* are, however, conservative because they guarantee that the regression surface on all $\mathbf{z}$-values are between the confidence bands. In many cases the set of possible linear combinations can be constrained somehow, and thus one gets narrower bands (e.g. Casella & Strawderman, 1980).

3.4.1 Confidence limits for the hazard in the logistic regression model

The above exact results rely upon the normality assumption. In the logistic model the distributional assumption leads to a different critical value distribution. Hauck (1983) and Piegorsch and Casella (1988) derived an approximate confidence band for the mean response $\mu = E(Y|\mathbf{Z}) = p(\mathbf{Z})$ of the logistic model. For our purposes this means, of course, just a confidence band for $p(t|\mathbf{Z})$ at each time point t; i.e., pointwise confidence intervals for $p(t|\mathbf{Z})$ at all values $\mathbf{Z} = \mathbf{z}$. In the following we consider these results as functions of time $t \in I$, where I is a fixed (discrete) time interval. Let $Z_t, t \in I$ be (possibly) random covariates. The mean $\mu(t|\mathbf{Z})$ can be modelled by a linear predictor $\eta(t) = \beta'\mathbf{Z}_{t-1}$, and these two quantities are linked via a link function g so that $g(\mu(t|\mathbf{Z})) = \eta(t)$. In this case the link function is the logit function, that is, $g(\mu(t|\mathbf{Z})) = ln(\mu(t|\mathbf{Z})/(1-\mu(t|\mathbf{Z})))$. The hazard $p(t|\mathbf{Z}) = \mu(t|\mathbf{Z})$ is related to the regression parameter β via the inverse link function $\mu(t|\mathbf{Z}) = g^{-1}(\eta(t)) = g^{-1}(\beta'\mathbf{Z}_{t-1}) = e^{\beta'\mathbf{Z}_{t-1}}/(1+e^{\beta'\mathbf{Z}_{t-1}})$.

Since g^{-1} is a monotonic function of $\beta'\mathbf{Z}_{t-1}$, asymptotic confidence limits for $p(t|\mathbf{Z}) = g^{-1}(\beta'\mathbf{Z}_{t-1})$ can be obtained by the inverse link as

$$g^{-1}(\beta'\mathbf{Z}_{t-1}) \in g^{-1}\left(\hat{\beta}'\mathbf{Z}_{t-1} \pm |c_{\alpha,p}|\sqrt{\mathbf{Z}_{t-1} I_{\hat{\beta}}^{-1} \mathbf{Z}_{t-1}}\right), \quad t \in I \qquad (3.30)$$

where $I_{\hat{\beta}}^{-1}$ is the estimated asymptotic covariance matrix of $\hat{\beta}$ as before. The confidence interval for $p(t|\mathbf{Z})$ at t is then

$$\left[(1+e^{-\hat{\beta}'\mathbf{Z}_{t-1}+|c_{\alpha,p}|\sqrt{\mathbf{Z}_{t-1}I_{\hat{\beta}}^{-1}\mathbf{Z}_{t-1}}})^{-1}, (1+e^{-\hat{\beta}'\mathbf{Z}_{t-1}-|c_{\alpha,p}|\sqrt{\mathbf{Z}_{t-1}I_{\hat{\beta}}^{-1}\mathbf{Z}_{t-1}}})^{-1}\right]. \quad (3.31)$$

In the usual Scheffé-method the critical value $c_{\alpha,p}^2$ is the $\chi_{p,\alpha}^2$-quantile since

$$(\beta - \hat{\beta})' I_{\hat{\beta}}^{-1} (\beta - \hat{\beta}) \sim \chi_p^2. \quad (3.32)$$

Piegorsch and Casella (1988) derived narrower confidence bands in the logistic model when some pre-specified constraints are put on the values of the covariates (e.g. indicator variables $Z_t = 0, 1$).

3.4.2 Stochastic order of failure time vectors

Let now $p_t^L(r)$ be the lower and $p_t^U(r)$ the upper endpoint of the confidence interval of the logistic hazard $p_t(r)$ of events $C_r, r = 1, ..., k$. Then, on the $(1-\alpha)$-confidence level, $p_t^L(r) \leq p_t(r) \leq p_t^U(r)$ holds for all t. We consider next conditions under which the monotonicity holds also for the prediction probabilities μ_t when they are based on functions of $p_t^L(r)$ and $p_t^U(r)$. For this purpose we need some concepts and results concerning *stochastic ordering*. A random variable S is *stochastically smaller* than a random variable S' (denoted by $S \underset{st}{\leq} S'$) if $P(S > u) \leq P(S' > u)$ for every u. Let $\mathbf{S} = (S_1, ..., S_k)$ and $\mathbf{S}' = (S_1', ..., S_k')$ be vectors of failure time variables. Then $\mathbf{S}$ is stochastically smaller than $\mathbf{S}'$ (denoted by $\mathbf{S} \underset{st}{\leq} \mathbf{S}'$) if $Eg(\mathbf{S}) \leq Eg(\mathbf{S}')$ for every bounded increasing real-valued function g.

We define the discrete time stochastic hazards as functions of the history process (H_t) as follows (cf. Chapter 2, (2.8))

$$p_t(r) = \sum_{n \geq 1} 1_{\{T_{n-1} < t \leq T_n\}} p^{(n)}(t, r | H_{T_{n-1}}), \quad r = 1, ..., k, \quad (3.33)$$

with the convention that $T_0 = 0$ and $H_{T_0} = \emptyset$.

Norros (1986) and Shaked and Shanthikumar (1987) have shown that when the compensators are absolutely continuous, an order relation between compensators based on histories satisfying a suitable order relation implies stochastic order of the corresponding failure time vectors. Using the 1-exponentiality of the cumulative hazards they construct a 1-1 correspondence between the failure time vector and a function of the 1-exponential variables. Norros (1986) calls this inverse mapping the *compensator representation* and Shaked and Shanhtikumar (1987) the *total hazard construction* of $\mathbf{S}$. The monotonicity of these functions implies stochastic ordering of the corresponding failure time vectors. This result is not directly useful in our discrete time model because no 1-1 correspondence exists in this case. Instead of that, we consider a result suggested by Arjas (private communication), which is based on a slightly stronger condition concerning the ordering of the local characteristics, the hazards $p_t(r)$.

For this purpose we need the concept of *supportivity* of a failure time vector (Norros, 1986). Intuitively, it means that in a system of components each failure increases the risk of the remaining components, the more the earlier the failure occurs. In Norros (1986), supportivity was defined in terms of the compensators.

We begin by defining an order relation between two history sets. Let H and H' be two history sets defined in (2.4). We write $H \leq H'$ if $(t,x) \in H'$ implies that $(s,x) \in H$ for some $s \leq t$. The interpretation is that the occurrences in H are earlier and more numerous.

Definition 3.1. Let $c(H) = \bigcup (r : \exists t : (t,r) \in H)$. The discrete failure time vector $\mathbf{S} = (S_1, ..., S_k)$ is called supportive if, for all r and t, $H \leq H'$ and $r \notin c(H)$ implies that $p_t(r|H) \geq p_t(r|H')$, $r=1,...,k$.

Our purpose is to show that the failure time vectors corresponding to the lower and upper endpoints of the confidence limits of $p_t(r)$ are stochastically ordered under the conditions of supportivity and hazard ordering; i.e.,

$$(S_1^L, ..., S_k^L) \underset{st}{\geq} (S_1^U, ..., S_k^U).$$

Since the following result is completely general, we denote the failure time vectors by $\mathbf{S}$ and $\mathbf{S}'$ and thereafter apply the result to the confidence limit setting.

Theorem 3.2. Let $\mathbf{S} = (S_1, ..., S_k)$ and $\mathbf{S}' = (S_1', ..., S_k')$ be two failure time vectors and $p_{t_j}(r|H)$ and $p'_{t_j}(r|H)$, $r = 1, ..., k$, the corresponding event-specific discrete time hazards at $t_j, j \geq 1$. Suppose

(i) $\mathbf{S}$ is supportive,

(ii) $H \leq H' \implies p_{t_j}(r|H) \geq p'_{t_j}(r|H')$ for all $j \geq 1$ and $r \notin c(H)$.

Then $\mathbf{S}' \underset{st}{\geq} \mathbf{S}$.

Proof. The proof is based on constructing two vectors $\hat{\mathbf{S}}$ and $\hat{\mathbf{S}}'$ of random variables which have the same probability distribution as $\mathbf{S}$ and $\mathbf{S}'$, and for which $\hat{\mathbf{S}}' \geq \hat{\mathbf{S}}$ holds in the sense of pointwise ordering. Consider timepoints $t_1 < t_2 < ...$ At each $t_j, j \geq 1$, and abbreviating $\hat{H}_{j-1} = \hat{H}_{t_{j-1}}$, we divide the unit (probability) interval into $k_j + 1$ segments, where k_j is the number of possible events at t_j, so that $c(\hat{H}_{j-1}) = k - k_j$. The lengths of the segments correspond to the given probabilities $p_{t_j}(r|\hat{H}_{j-1}), r \notin c(\hat{H}_{j-1})$, and the length of the $(k_j + 1)$st segment is $1 - \sum_{i: i \notin c(\hat{H}_{j-1})} p_{t_j}(i|\hat{H}_{j-1})$. By (ii), we have within each segment a "subsegment" of length $p'_{t_j}(r|\hat{H}_{j-1}) \leq p_{t_j}(r|\hat{H}_{j-1})$. If $r \in c(\hat{H}_{j-1})$ for S_r, then the length of the rth segment is $p'_{t_j}(r|\hat{H}_{j-1})$.

At $t_j, j \geq 1$, let Z_j be a uniform (0,1) random variable. Depending on the value of Z_j we have three possibilities:

(1) if $\sum_{\substack{i=1 \\ i \notin c(\hat{H}_{j-1})}}^{r-1} p_{t_j}(i) + p'_{t_j}(r) < Z_j \leq \sum_{\substack{i=1 \\ i \notin c(\hat{H}_{j-1})}}^{r} p_{t_j}(i)$, for some $r \notin c(\hat{H}_{j-1})$, then
$\hat{S}_r = t_j$ and $\hat{S}'_r > t_j$.

(2) if $\sum_{\substack{i=1 \\ i \notin c(\hat{H}_{j-1})}}^{r-1} p_{t_j}(i) < Z_j \leq \sum_{\substack{i=1 \\ i \notin c(\hat{H}_{j-1})}}^{r-1} p_{t_j}(i) + p'_{t_j}(r)$, for some $r \notin c(\hat{H}_{j-1})$, then
$\hat{S}'_r = \hat{S}_r = t_j$.

(3) if $Z_j > \sum_{\substack{i=1 \\ i \notin c(\hat{H}_{j-1})}}^{k_j} p_{t_j}(i)$ then $\hat{S}'_r > t_j$ and $\hat{S}_r > t_j$ for all $r \notin c(\hat{H}_{j-1})$.

This construction, performed at each $t_j, j \geq 1$, yields two random vectors $\hat{\mathbf{S}}$ and $\hat{\mathbf{S}}'$. The pointwise inequality $\hat{\mathbf{S}}' \geq \hat{\mathbf{S}}$ follows now by showing that the corresponding histories $\hat{H}_j$ and $\hat{H}'_j$ satisfy $\hat{H}_j \leq \hat{H}'_j$ for all j. The proof is by induction on j. Obviously $\hat{H}_0 = \hat{H}'_0 = \emptyset$. Suppose that $\hat{H}_{j-1} \leq \hat{H}'_{j-1}$. Then by (i) and (ii),

$$p_{t_j}(r|\hat{H}_{j-1}) \underset{(i)}{\geq} p_{t_j}(r|\hat{H}'_{j-1}) \underset{(ii)}{\geq} p'_{t_j}(r|\hat{H}'_{j-1}), \text{ for all } r \notin c(\hat{H}_{j-1})$$

It is a direct consequence of (1)-(3) above that also $\hat{H}_j \leq \hat{H}'_j$. Since by construction $\hat{\mathbf{S}} \underset{st}{=} \mathbf{S}$ and $\hat{\mathbf{S}}' \underset{st}{=} \mathbf{S}'$ we have $\mathbf{S}' \underset{st}{\geq} \mathbf{S}$.

We now return to the confidence limits of the prediction probabilities. Recall the causal chain model in Chapter 2 for the consecutive events $C_1, C_2, ..., C_k$. We apply Theorem 3.2 to prove that supportivity and local dominance imply stochastic ordering of the occurrence times, and thus monotonicity of the prediction probabilities as functions of the histories.

Theorem 3.3 *Let C be an (ordered) k-point causal chain. Let $p_t^U(r)$ and $p_t^L(r)$ be the discrete time upper and lower endpoints of the confidence limit of the hazard of event C_r, $r = 1, ..., k$. Let $\mathbf{S}^U = (S_1^U, S_2^U, ..., S_k^U)$ and $\mathbf{S}^L = (S_1^L, S_2^L, ..., S_k^L)$ be the corresponding occurrence time vectors. Suppose $\mathbf{S}^U$ is supportive. Then $\mathbf{S}^U \underset{st}{\leq} \mathbf{S}^L$.*

Proof. The result follows from Theorem 3.2, where in condition (ii) $p_{t_j}^U(r) = p_{t_j}(r)$ and $p_{t_j}^L(r) = p'_{t_j}(r), r = 1, ..., k$, for which by definition $p_{t_j}^L(r) \leq p_{t_j}^U(r)$ for all j, $r = 1, ..., k$.

A stronger result for supportivity is based directly on the ordering of the prediction processes as functions of the history. A system of components is called *strongly supportive* if, for all t, $H \leq H'$ implies $\mu_t^*(H; I) \geq \mu_t^*(H'; I)$ (Arjas & Norros, 1990). For practical purposes the hazard ordering of Def. 3.1. provides, however, a relatively easy way to establish the monotonicity of the prediction probabilities, which was our purpose when considering their confidence limits.

3.5 Discussion

The results concerning supportivity are useful in the (special) case in which the influence of all events is *causative* (positive) on all subsequent events, that is, their occurrence increases the hazard of the next event(s). Such systems are natural in reliability theory in which a system's functioning is dependent on the state of its components which can be either on or off. In applications that we have in mind here, some events can be preventive (with respect to the response), or causative, but in a way that the later the event occurs the more it increases the risk of the response. Such chains are not supportive. This is the case in both examples of Chapter 4.

The generalized delta-method, presented in Section 3.2, is applicable also in other situations. In parametric models, such as the logistic regression model, the analytic derivatives of the prediction probabilities exist and they can be obtained while calculating the prediction probabilities. However, it is well-known that because the delta-method is restricted to the first-order approximation of the likelihood function these so called Wald confidence regions can sometimes be rather misleading if the true confidence region is far from an ellipsoid. If the number of parameters p is small, preferably $p \leq 2$, it is a reasonable task to calculate the likelihood-based confidence regions by determining the profile likelihood of each parameter and the joint confidence region can be obtained graphically. In our case the number of parameters in all hazard regression models can easily extends 20. Matthews (1988) has suggested a method of finding solutions for constrained likelihoods to obtain approximate likelihood-based confidence limits for functions of several parameters. In his examples from survival analysis, over 100 non-parametric ML-estimates were needed, depending on the number of time intervals. However, Keiding and Andersen (1989) received, when using one of his example data, very similar results by calculating the Wald-type regions.

Even though the analytic derivatives are available in the logistic model, the calculations become formidable if the number of events k is large, say $k > 4$, or if there are several response events as in the first example of Chapter 4. In general, if the follow-up is long and the time unit small, and in addition the number of parameters p is large, the matrices in the quadratic forms of the asymptotic covariances become very soon extremely large. This part of the analysis is clearly the most elaborate, and more efficient methods are needed.

4. APPLICATIONS

4.1 Multistate models in follow-up studies

The dynamic causal chain considered in Chapter 2 is a special case of a *multistate* survival model which generalizes the simple two-state "alive-dead" survival model. The assumption of ordered occurrence times restricts the events to be nonrecurrent. There is of course no mathematical reason for it. It is made only for the interpretation of the causal structure of the events. In practice it is sometimes sensible to let the order of the events to be undetermined in advance, although in our opinion this does not necessarily correspond to causal analysis anymore (cf. the application of this method in Klein *et al.*, 1993).

Multistate models have been used for a long time in biostatistics (e.g. Chiang, 1968, Prentice *et al.*, 1978). In demography and in the actuarial science models with multiple causes of death are traditionally called *multiple-decrement* models (e.g. Elandt-Johnson & Johnson, 1980) referring to life tables with multiple end points. In modern survival analysis these kind of data, the *event-histories*, are considered as realizations of stochastic processes.

In medical studies multistate models have been used to model, for example, the *progression of a disease* through transient states so that the "alive" state is split into two or more transient states, and the "dead" state is absorbing. The simplest model is the 3-state "illness-death" model which has proved to be useful in studying the differential response (state 3) of patients diagnosed to have a particular disease (state 1), some of which subsequently develop an additional complication (state 2). Some examples are the effect of graft-versus-host disease on relapse among leukemia patients after bone marrow transplantation (e.g. Weiden *et al.*, 1981, Pepe *et al.*, 1991, Pepe, 1991), and the effect of nephropathy on mortality among diabetic patients (Andersen, 1988). In cancer studies so called multistage or multihit models have been used for a long time to model the transition of normal cells into malignant cells (e.g. Armitage & Doll, 1954, Moolgavkar & Knudson, 1981). Recently the dynamics of AIDS (e.g. Longini *et al.*, 1989) has been studied by using multistate models.

Another generalization of the two-state survival model, which is also considered in one of our examples, is the *competing* risks model. There we instead of several transient or intervening states have several alternative response states. If the alternative responses are dependent in the sense that the occurrence of one of them hinders the other's occurrence (e.g. death from one cause prevents from others), the marginal survival functions are not estimable. However, the probability distributions and cumulative hazards of each alternative response can always be estimated.

Usually the simplifying assumption has been made that the multistate models are Markovian, or sometimes semi-Markovian. The well established theory of Markov processes is then available. Aalen (1976, 1978) first considered non-homogenous Markov models in the framework of counting processes, and later Aalen and Johansen (1978) provided nonparametric estimators for the transition probabilities and their asymptotic variances and covariances using martingale methods. The results have since then used, for example, by Keiding and Andersen (1989) and Andersen et al. (1991). Pepe (1991) considered the asymptotic properties of functions of several non-parametric statistics, such as cumulative hazard, incidence function and survival function, in multiple endpoint models. Even though it is often recognized that the primary interest in multistate models is actually in the (survival) probabilities and not in the conditional (instantaneous) hazards, rather few authors have considered the statistical properties of the probabilities when incorporating covariate information. As an exception, Andersen et al. (1991) use the results in Aalen and Johansen (1978) to derive asymptotic covariances for the transition probabilities in the semi-parametric Cox model with a Markov assumption.

4.2 Modelling dependence between causal events

Hume's localization principle, which was presented in Chapter 1, implies that causal events must be contiguous in time. As mentioned, in causal chains this can be interpreted as the requirement that the chain is Markovian. When this restriction is relaxed, various forms of dependencies between the causal events can be suggested. We consider next some possible functional forms and for convenience illustrate them in a 'Cox-type' *multiplicative hazard* model with exponential risk function; i.e., all transformations are considered in the additional exponential transformation on the hazard.

In Chapter 2 we already presented a hypothetical example of an additive hazard model in which the influence of the causes were *cumulative*. This implies that earlier occurrences of a cause C are more hazardous than later ones since the causal effect of C on E is cumulating in time. Let now $T_C = v$. Then the conditional hazard of E, given C at v, is in a multiplicative model of the form

$$\lambda_{E|C}(t|v) = \lambda_E(t) exp(\alpha + \beta(t-v)), \ t \geq v, \qquad (4.1)$$

and consequently $\lambda_{E|C}(t|v) \geq \lambda_{E|C}(t|v')$, if $v \leq v'$ and $\beta > 0$. If the data suggest a more moderate increase for large intervals between C and E, a more appropriate functional form would be to replace $t - v$ by $log(t - v)$.

The influence of C on the hazard of E can on the other hand be *transient* so that the hazard of E goes up right after $t = T_C$ but decreases then gradually, and stays ultimately on a level higher than without the occurrence of C. In the multiplicative hazard model this would be satisfied if

$$\lambda_{E|C}(t|v) = \lambda_E(t)exp(\alpha + \beta e^{-\gamma(t-v)}), \ t \geq v, \qquad (4.2)$$

where $\alpha > 0$ is a parameter of the level of the conditional hazard $\lambda_{E|C}(t|v)$ irrespective of t, $\beta > 0$ is a parameter of the level and slope of the exponential curve, and $\gamma > 0$ is the rate of transience.

Sometimes the time between C and E is not informative but the hypothesis is that the actual occurrence time of C has differential effect on the hazard of E, for example, in the way that an earlier occurrence of C raises the hazard more than a later one (or vice versa). Then it is expected that in a population the number of responses E is more numerous for earlier occurrences of C. The hazard $\lambda_{E|C}(t|v)$ is of the same form as $\lambda_E(t)$ but starts at $T_C = v$ on a higher (lower) level α, and the size of the jump depends on v, i.e.,

$$\lambda_{E|C}(t|v) = \lambda_E(t)exp(\alpha + \beta v), \ t \geq v. \qquad (4.3)$$

Several transformations to weigh the influence of earlier occurrences of C can be suggested. For example, if a more rapid increase of the hazard is expected when C occurs early then transformations like $\sqrt{T_C}$ or $log(T_C)$ could be used, or if the hazard is expected to decrease when C occurs late, then $1/T_C$ or $1/\sqrt{T_C}$ could be appropriate transformations.

A crude way to model *non-monotonic* influence of T_C on the hazard of E is to define the hazards as step functions $\sum_k 1_{\{c_{k-1} < T_C \leq t \leq c_k\}} \cdot a_k$, depending on the index of the pre-specified intervals. Furthermore, a "U-shaped" second order influence (in the exponential risk function) of T_C on the hazard can be modelled by

$$\lambda_{E|C}(t|v) = \lambda_E(t)exp(\alpha + \beta(v-c)^2), \ t \geq v. \qquad (4.4)$$

where $c > 0$ is some constant.

Finally, there may be some (perhaps logical) *causal lag* or "reaction time" d between the occurrences of C and E such that $\lambda_{E|C}(t|v) = \lambda_E(t)$ until $T_C + d, d > 0$, and then $\lambda_{E|C}(t|v)$ jumps to a level α; i.e.,

$$\lambda_{E|C}(t|v) = \lambda_E(t)exp(1_{\{T_C + d \leq t\}}(\alpha + f(t,v))), t \geq v \qquad (4.5)$$

where the function $f(t,v)$ can now be either 0 or, for example, some of the above functions (e.g., exponential decay) of t and v. If $f = 0$, a cause C has a *constant* effect on the hazard of E regardless of T_C. The above cases with different forms of f are of course just special cases of this general formula with $d = 0$.

In real data sets such simplified hypothetical forms of dependencies sometimes fit only for subsets of the data (for example, for some age groups). Obviously one also needs a considerable amount of data to estimate such complicated time-dependencies. Even in large data sets the events of interest are sometimes so rare that such estimation is impossible since the parameters of these terms are estimated among those who actually experienced the causes. The requirements for the quality of the data are also quite stringent; exact dates on all causal events are needed on all observations. These problems also concern the two real data examples in 4.3.

4.3 Two applications

The next two data sets have both been analysed previously by ordinary hazard methods, and we reanalyze them in order to test the usefulness of the prediction method. They also serve as examples of the limitations of this method. The first example is a typical clinical follow-up study in which the failure of a treatment is studied in a rather small number (n=163) of patients. The structure of the causes is quite complicated since the failure can be either death due to complications following the treatment, or a new illness episode, relapse. Some of the causes have opposite effects on these two response events. The interest is therefore in investigating how the predictions of different patients (i.e., patients with different histories at time t) change after the occurrence of a particular cause, and what is then the "balance" between the two competing modes of failure.

The second example is a large sample (n=6789) observational study in which a group of children in a particular area has been followed from birth until a certain age. For an observational study it has an ideal setting for causal analysis since the *observational plan is independent of the causal events*; all births in this particular area are included, and it is not known which children are going to be exposed or diseased later on. This differs from the observational plan in which the sample consists of those exposed or, in a retrospective study, of those diseased. The subjects are therefore not selected to the study in a way that depends on the causes or on the response. This is perhaps as close as we can get to when imitating a randomized experiment in observational studies. Hoem (1986) and Kalbfleisch and Lawless (1988) call such an observational plan *noninformative*. It also closely corresponds to the notion of an *ignorable* sample plan of Rubin and Rosenbaum (1983).

4.3.1 The Nordic bone marrow transplantation data: the effects of CMV-infection and chronic GvHD on leukemia relapse and death

In a Nordic multicenter study 163 patients with acute myeloid leukemia (AML) or acute lymphoplastic leukemia (ALL) were treated by allogeneic bone marrow transplantation between the years 1980 and 1985, and the recurrence of acute leukemia (relapse) or death were registered. In several studies (e.g. Weiden *et al.*, 1979, Zwaan *et al.*, 1984) it has been found that chronic graft-versus-host disease (GvHD), which is a donor mediated systemic inflammatory reaction, and itself causes severe morbidity, has antileukemic effect; i.e., it decreases the risk of relapse. The mechanism that mediates this antileukemic effect, and the properties of the donor which predict reduced risk of relapse are, however, not known. Recently, Lönnqvist et al. (1986) showed that patients with post-transplant cytomegalovirus (CMV) infection had a reduced risk of relapse. In Jacobsen et al. (1990) the CMV-immune status of the donor and the probability of staying in remission after transplantation were studied by hypothesizing that the antileukemic effect is transmitted by immunologic reactions of the transplanted cells.

In Jacobsen et al. (1990) the relationships between CMV-infection, chronic GvHD, relapse and death were analysed by traditional survival methods; the probability of staying in remission after the bone marrow transplantation (BMT) was estimated nonparametrically by the Kaplan-Meier estimator and the effects of the risk factors on the hazard of chronic GvHD, relapse and death were estimated by separate Cox regression models. The thorough analysis in Jacobsen et al. (1990) gives a good picture of the dependencies between the events, and we shall use the same covariates as in the Cox models in our hazard models.

4.3.1.1 The causal chain

A preliminary analysis by the Nelson-Aalen plots in Fig.4.1 and smoothed hazard functions in Fig.4.2 shows that the most critical period after transplantation are the first 3-4 months. Almost half of the patients who died without relapse died during the first 100 days due to complications. All four events occur most frequently in the first 200 days. According to the conventional definition of acute and chronic GvHD, symptoms which appear before about 90 days since transplantation are called acute, and those after about 90 days chronic, although the same patient can have both acute and chronic GvHD subsequently. Since acute GvHD occurred almost entirely before 50 days, we use it just as an indicator (i.e., whether a patient had acute symptoms of GvHD or not), and calculate separate predictions for each case. This also simplifies the chain of events considered. Fig.4.1 shows that the occurrence of CMV-infection is most common before 90 days after transplantation, and since chronic GvHD does not occur until 90 days by definition, this suggests a natural order for these two events in the chain although there is no known obvious biological explanation for it.

In this example we consider two competing response events, relapse and death, which are both indicators of failure of the treatment. The results in Jacobsen et al. (1990) indicate that acute GvHD and CMV-infection increase the hazard of death in remission whereas they both decrease the hazard of relapse. Chronic GvHD protects against relapse but, at least in these data, it seems to have no effect on death in remission. This suggests that analysing the two responses as competing failures rather than separately might give more insight into the complicated causal structure. We therefore consider the following causal chain

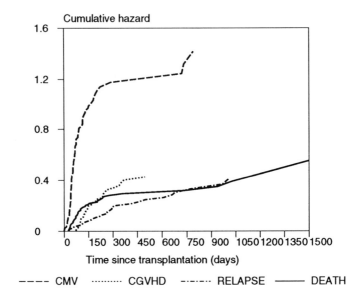

Fig. 4.1. Estimated cumulative hazards of CMV-infection, chronic GvHD, relapse and death in remission in the BMT data

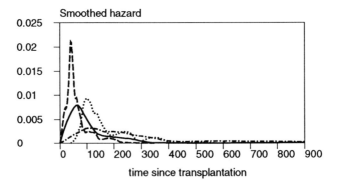

Fig. 4.2. Smoothed hazard functions of CMV-infection, chronic GvHD, relapse and death in remission in the BMT data

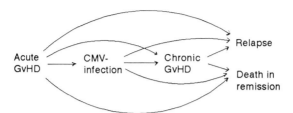

and denote the corresponding occurrence times by T_C, T_G and T_R/T_D.

Obviously a patient can die after relapse but in this setting these cases are treated as relapses. Thus, a patient can at time t either die in remission, relapse or be alive relapse-free. Even though knowing the probabilities of two of the three possible events at some t automatically gives the probability of the third at that time, it is of interest to investigate how different histories at the time of prediction affect these three probabilities.

4.3.1.2 The data

A detailed description of the data and the treatment procedure to detect acute and chronic GvHD and CMV-infection after transplantation is given in Jacobsen *et al.* (1990). At the time of transplantation the patient age ranged from 1 to 49 years and 98 of the 163 patients were under 20 years. In Table 4.1 the diagnosis and stage of the disease are shown together with donor age and donor CMV-status. Table 4.2 shows the patient's post-transplant status in certain important risk factors.

A year after transplantation the risk of both relapse and death decreases rapidly, and since by definition $T_C \geq 90$ days, we restrict the prediction interval to $(90, 360]$ in order to be able to compare the predictions of all possible histories, including those in which chronic GvHD has already occurred.

4.3.1.3 The hazard models

The same hazard models as in Jacobsen *et al.* (1990) were used for the four events except that each preceding event in the chain was included in the next hazard model even if its coefficient was not significant. This is the case for CMV-infection and chronic GvHD since even though acute GvHD increases the risk of CMV-infection there seems no significant link between chronic GvHD and CMV-infection. Another difference compared to the Cox models in Jacobsen *et al.* (1990) is that the underlying time is also parameterized. However, we used a grid value of one day, so actually the continuous time-scale was approximated as closely as one can. The Nelson-Aalen plots and the smoothed hazard functions in Figs. 4.1 and 4.2 were used to define the cutpoints of the underlying time variable t. They are modelled as indicator variables.

Table 4.1 Patient and donor characteristics

Diagnosis and stage of disease	n	Donor age		Donor CMV status	
		<20	>20	Sero-	Sero+
Acute ML	71	25	46	28	43
Acute LL	92	57	35	49	43
Induction failure	6	5	1	3	3
First remission	69	30	39	30	39
2nd-4th remission	72	42	30	34	38
1st-3rd relapse	16	5	11	10	6
Total	163	82	81	77	86

Table 4.2 Absolute numbers of relapse and death in the presence of risk factors

Risk factor	n	Post-transplant event	
		Relapse	Death in remission
Donor < 20 years	82	19	15
Donor > 20 years	81	17	30
Donor CMV-neg.	77	27	16
Donor CMV-pos.	86	9	29
Acute GvHD present	58	8	33
Acute GvHD absent	105	28	12
Chronic GvHD present	37	2	9
Chronic GvHD absent	126	34	36
CMV infection present	106	15	36
CMV infection absent	57	21	9
Total	163	36	45

(Source: Jacobsen *et al.* 1990)

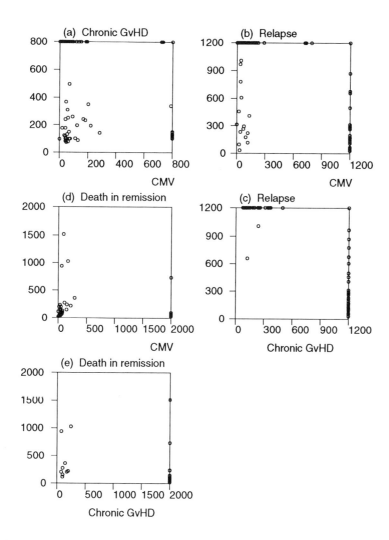

Fig. 4.3. Plots of the occurrence times of (a) chronic GvHD vs. CMV-infection (b) relapse vs. CMV-infection (c) relapse vs. chronic GvHD (d) death in remission vs.CMV-infection (e) death in remission vs. chronic GvHD.

In order to get some insight into the dependencies between the events we plotted the occurrence times of CMV-infection and chronic GvHD against the corresponding response times of relapse and death (Fig. 4.3). The values on the axes are those of the censored cases; i.e., those who did not experience the considered events. It is evident from Fig. 4.3 that the small number of cases with preceding causal event does not allow to estimate complicated time-dependencies between the events. The effect of the preceding events is therefore modelled only as a time-dependent indicator. This corresponds to the assumption that the influence of a cause is the same regardless of when it occurs, and all that matters is whether it has occurred by t, or not. This has an obvious effect on the 3-phase analysis of the prediction probabilities where we fix t, u or H one at a time. We shall return to that below. Before going to the analysis of the prediction probabilities we briefly comment on the risk factors in each hazard model. The complete results of the hazard analyses are in Appendix 2.

CMV-infection. Three patient characteristics: positive *CMV-immune status* (CMV-=0, CMV+=1), *mismatch* between patient and donor cells (0=no, 1=yes), and old *donor age* (0=under 20 years, 1=over 20 years) together with acute GvHD, which was estimated as a time-dependent indicator, increased the risk of CMV-infection. The model is the following (the indicator function $1_{\{\cdot\}} = 1$, if the condition $\{\cdot\}$ is true, and 0 otherwise):

$$ln(\frac{P(\Delta N_t(C) = 1|\mathcal{F}_{t-1})}{P(\Delta N_t(C) = 0|\mathcal{F}_{t-1})}) = \alpha_0 + \alpha_1 1_{\{PCMV=1\}} + \alpha_2 1_{\{MISM=1\}}$$
$$+ \alpha_3 1_{\{DAGE=1\}}$$
$$+ \alpha_4 1_{\{T_A \leq t\}} + \alpha_5 1_{\{t \geq 90\}}$$

Chronic GvHD. Old *donor age* (> 20 years) and other than *female to male donation* (0=no, 1=yes) together with preceding acute GvHD increase the hazard of chronic GvHD, whereas preceding CMV-infection was not an important predictor for chronic GvHD. The model for chronic GvHD is

$$ln(\frac{P(\Delta N_t(G) = 1|\mathcal{F}_{t-1})}{P(\Delta N_t(G) = 0|\mathcal{F}_{t-1})}) = \beta_0 + \beta_1 1_{\{DAGE=1\}} + \beta_2 1_{\{F/M=1\}}$$
$$+ \beta_3 1_{\{T_A \leq t\}} + \beta_4 1_{\{T_C \leq t\}}$$
$$+ \beta_5 1_{\{0 \leq t < 150\}} + \beta_6 1_{\{150 \leq t < 250\}}$$

Relapse. The most significant predictors for relapse were the *stage of the disease* when transplanted and old *donor age* (> 20 years). The stage of the disease was measured by two variables: transplantation after 1. remission and transplantation under relapse. For a patient who was operated during the second relapse, the intensity factor (e^β) is then a product of the two factors. Late stage of the disease increases the risk of relapse most, and therefore a patient who has experienced several relapses already has the highest hazard for a new relapse. All preceding events, acute GvHD, CMV-infection and chronic GvHD, are protective;

CMV-infection, however, only among those with seropositive CMV-status. The model is

$$\ln\left(\frac{P(\Delta N_t(R) = 1|\mathcal{F}_{t-1})}{P(\Delta N_t(R) = 0|\mathcal{F}_{t-1})}\right) = \gamma_0 + \gamma_1 1_{\{UREL=1\}} + \gamma_2 1_{\{A1REM=1\}}$$
$$+ \gamma_3 1_{\{DAGE=1\}}$$
$$+ \gamma_4 1_{\{T_A \leq t\}} + \gamma_5 1_{\{T_G \leq t, PCMV=1\}} + \gamma_6 1_{\{T_C \leq t\}}$$
$$+ \gamma_7 1_{\{250 \leq t < 600\}} + \gamma_8 1_{\{t \geq 600\}}$$

Death in remission. The same predictors as for relapse were significant also for the death hazard. Here the preceding events have, however, different importance than for relapse, as was pointed out earlier. Acute GvHD and CMV-infection (regardless of CMV-status) increase the risk of death unlike the risk of relapse. The model for death in remission is

$$\ln\left(\frac{P(\Delta N_t(D) = 1|\mathcal{F}_{t-1})}{P(\Delta N_t(D) = 0|\mathcal{F}_{t-1})}\right) = \delta_0 + \delta_1 1_{\{UREL=1\}} + \delta_2 1_{\{A1REM=1\}}$$
$$+ \delta_3 1_{\{DAGE=1\}}$$
$$+ \delta_4 1_{\{T_A \leq t\}} + \delta_5 1_{\{T_G \leq t\}} + \delta_6 1_{\{T_C \leq t\}}$$
$$+ \delta_7 \log(t)$$

4.3.1.4 The prediction probabilities

Since relapse and death in remission are competing failures we denote by $S = T_R \wedge T_D$ the time at which either relapse or death occurs first. Then, using the notation in Chapter 2, we denote the discrete time subdistribution of R (resp. D) for the interval $(t, u]$ by

$$P(t < S \leq u, X = R|\mathcal{F}_t) = F_R(u|H_t). \quad (4.6)$$

The sum
$$F(u|H_t) = F_R(u|H_t) + F_D(u|H_t) \quad (4.7)$$

is the probability that either one occurs before u. Consequently, the probability of relapse-free survival at least until u is

$$P(S > u|\mathcal{F}_t) = 1 - F(u|H_t). \quad (4.8)$$

Note that the marginal survival functions of relapse or death in remission are not estimable since the competing response events are not independent. The prediction probabilities of relapse-free survival, relapse and death in remission are of the same form as those in Chapter 3 with four possible histories at t.

4.3.1.5 Fixed starting point t, fixed history H, endpoint u varies

We start by showing alternative predictions of the three possible response states: relapse-free survival, relapse and death in remission when the endpoint u varies within $(90, 360]$ days after transplantation. The possible histories at the time of prediction are $\{(v, C)\}$, $\{(w, G)\}$, $\{(v, C),(w, G)\}$, $v \leq w \leq t$, and $\{\emptyset\}$. Two sets of predictions were calculated; the one with no acute GvHD before t (Figure 4.4) and the other with acute GvHD before t (Figure 4.5). In the figures we consider a hypothetical patient with the following covariates: patient CMV+, no mismatch, donor age > 20 years, female to male donation, and transplantation after first remission. Different values of the fixed covariates give of course different predictions, but the shape of the curves is similar.

First of all, it is clear from Figs. 4.4 and 4.5 that both CMV-infection and chronic GvHD protect from relapse and that CMV-infection increases the probability of death in remission within the one year prediction interval. Since the probability of relapse-free survival depends on the balance between the risks of relapse and death in remission, we consider the effect of acute GvHD on these two responses first. Acute GvHD has stronger influence on death in remission which is natural because death in remission is usually a consequence of complications due to transplantation and acute graft-versus-host disease within three months after transplantation probably reinforces the effect of these complications. The influence of acute GvHD is very strong; the prediction probabilities of death in remission before $u = 360$ as functions of different histories are 6- to 9-fold when acute GvHD has occurred before t compared to no preceding acute GvHD. If both acute GvHD and CMV-infection occur ($T_C = 90$), the probability of death in remission is highest among all possible histories (change from 0.11 to 0.65 if acute GvHD is present).

In contrast, chronic GvHD has a slightly protective effect on death in remission. This might be due to the fact that those who develop chronic GvHD must have survived at least three months after transplantation, which is, as already noted, the critical period for complications resulting in death. Although all the predictions considered here start from $t = 90$, the coefficient of chronic GvHD in the death hazard model is estimated among those still alive at $t = 90$. Even though acute GvHD increases the risk of death and decreases that of relapse, the two probabilities of relapse-free survival, $P(T_R > t, T_D > t|A)$ and $P(T_R > t, T_D > t|\emptyset)$, are almost of the same size because in the considered prediction interval the risk of death in remission is much higher than that of relapse. This is an example of how difficult it is to tell in advance what the "balance" between the competing failures can be, even though we know that acute GvHD has a strong opposite influence on them.

When predicting relapse before $u = 360$ at $t = 90$, acute GvHD alone decreases the probability of relapse to a half (from 0.48 to 0.22 in the $\emptyset$-history). If additionally CMV-infection occurs (at $T_C = 90$), the probability of relapse becomes almost negligible (decrease from 0.12 to 0.03). Chronic GvHD already alone prevents from relapse effectively, and the additional effect of preceding acute GvHD is small (from 0.08 to 0.04 when acute GvHD present). When both CMV-infection

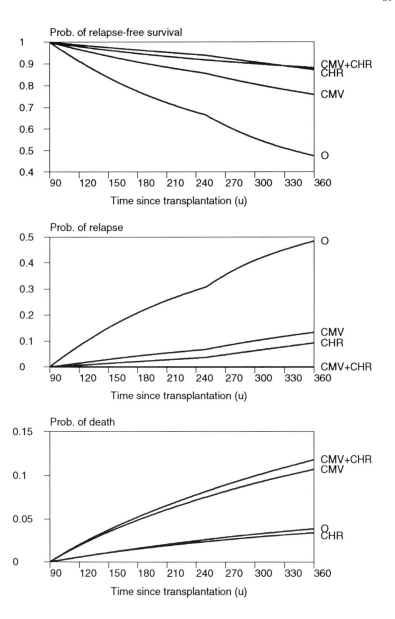

Fig. 4.4. Probability of (a) relapse-free survival, cumulative probabilities of (b) relapse and (c) death in remission, evaluated at $t = 90$ days after transplantation, for alternative histories: CMV=CMV-infection, CHR=chronic GvHD, O=neither at t (no acute GvHD before t).

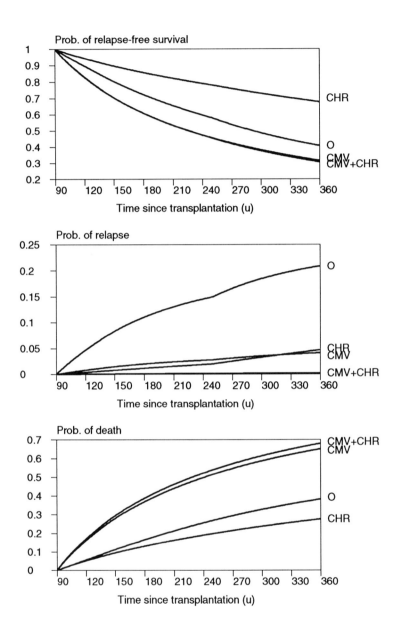

Fig. 4.5. Probability of (a) relapse-free survival, cumulative probabilities of (b) relapse and (c) death in remission before u, evaluated at $t = 90$ days after transplantation, for alternative histories: CMV=CMV-infection, CHR=chronic GvHD, CMV+CHR= both, O=neither at t (acute GvHD before t).

and chronic GvHD have occurred ($T_G = 90$), the probability of relapse is negligible, regardless of the occurrence of acute GvHD.

4.3.1.6 Fixed history H, fixed endpoint u, starting point t varies

Consider now the function $t \mapsto \mu_t^*(H_t; (90, 360])$. Choosing $t = T_C$ or T_G makes no difference because the influences of CMV-infection and chronic GvHD were modelled as indicators (i.e., constant effect). Therefore the changes in the probabilities (Figs. 4.6 and 4.7) and in the innovation gains (Figs. 4.8-4.11) are purely due to the varying length of the prediction interval (as t varies) and the changing probabilities of relapse and death in remission within the prediction interval. The main information therefore concerns the size of the innovation gains as functions of different histories.

We consider the innovation gains of CMV-infection, chronic GvHD, and chronic GvHD given CMV-infection previously, again separately for relapse-free survival, relapse and death. The hazard of death in remission decreases very rapidly after transplantation, and therefore the prediction of dying in remission within the one year prediction interval decreases the later the prediction starts before u. This concerns both short term (u=180) and long-term (u=360) predictions of death whereas the long-term prediction of relapse is almost constant regardless of the occurrence time of either CMV-infection or chronic GvHD. All innovation gains of the short-term predictions (u=180) tend to zero near u (and all prediction probabilities to 1 at u), which is just a consequence of the fact that the occurrence of the causal event at u does not change the probability of the response at u.

The innovation gains from chronic GvHD (alone) for relapse are negative, indicating a protective effect as large as 40% when the prediction extends to one year (u=360) after transplantation, whereas those for death in remission are negligible, and tend to zero both in short-term and long-term predictions when $t \to 180$. The innovation gains from CMV-infection for relapse are as large as those from chronic GvHD alone, whereas for death in remission they are positive, indicating that CMV-infection increases the risk of death in remission. They become smaller also in the long-term prediction (u=360) because the probability of death decreases rapidly after 100 days. Again, the opposite influence of the causes on the competing failures means that even though the innovation gains of CMV-infection and chronic GvHD for relapse (without preceding acute GvHD) are almost equal, the positive effect of chronic GvHD is larger on relapse-free survival because CMV-infection increases the risk of death.

We compare next the innovation gains from chronic GvHD *given CMV-infection any time previously*. Again, we consider the predictions without and with preceding acute GvHD (Figs. 4.10 and 4.11) separately. Without acute GvHD the innovation gains from chronic GvHD for relapse are now about 30% of those from chronic GvHD alone (cf. Figure 4.9 (b)), and only 10% of those where acute GvHD was preceding chronic GvHD (cf. Fig. 4.11 (b)). So, the "new" information and protective effect of chronic GvHD is rather modest if we already know that CMV-infection and, especially both acute GvHD and CMV-infection, have

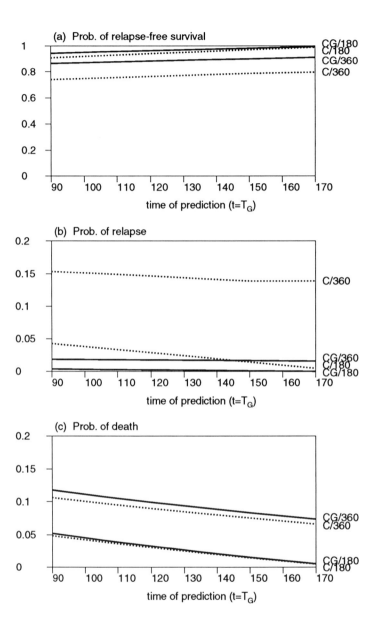

Fig. 4.6. Probability of (a) relapse-free survival, cumulative probabilities of (b) relapse and (c) death in remission before u, when CMV-infection has occurred any time previously and chronic GvHD has just occurred (CG), and when chronic GvHD has not yet occurred (C) (no acute GvHD before t).

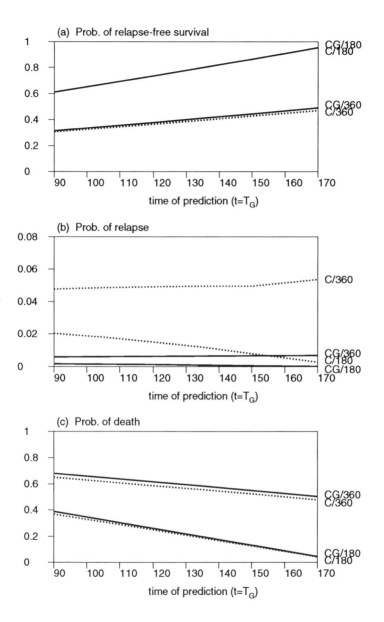

Fig. 4.7. Probability of (a) relapse-free survival, cumulative probabilities of (b) relapse and (c) death in remission until u, when CMV-infection has occurred any time previously and chronic GvHD has just occurred (CG), and when chronic GvHD has not yet occurred (C) (acute GvHD before t).

occurred previously. Interestingly, the innovation gains from chronic GvHD for death in remission, given CMV-infection previously, are positive, indicating that the occurrence of chronic GvHD in that situation increases the risk of death, although very little. This happens regardless of preceding acute GvHD, but the effect is larger when also acute GvHD has occurred. When considering relapse-free survival, the innovation gains from chronic GvHD (without acute GvHD) are less than one third of those without CMV-infection (cf. Fig 4.9 (a)), and become negligible when also acute GvHD precedes chronic GvHD.

We also considered models where the interactions of the events were taken into account (i.e., $1_{\{T_C \leq t\}} \times 1_{\{T_G \leq t\}}$). These interaction terms were, however, not significant (in the logistic model). One explanation for this is, of course, the small number of relapses and deaths with preceding causal events.

The asymptotic 95% pointwise confidence limits were calculated for the innovation gains from observing chronic GvHD when CMV-infection, but not acute GvHD, had already occurred (Fig. 4.12). Note that in all figures the limits for the short term prediction tend to zero when $t \to u$. This illustrates again the fact that no matter if chronic GvHD occurs at u or not, the probability that relapse or death occurs at u is zero, if they have not occurred before. The limits for the innovation gains in predicting relapse are very narrow even for the long-term predictions, indicating very strongly that even when CMV-infection has occurred previously, chronic GvHD still decreases the probability of relapse before $u = 360$ about 11-12%. In contrast, the causal effect of chronic GvHD on death in remission or consequently, on relapse-free survival is rather uncertain, ranging for relapse-free survival between 6-15% ($u = 360$), and for death in remission even the value of 0 is included in the interval.

4.3.1.7 Fixed starting point t, fixed endpoint u, history H varies

In the constant effects model this alternative gives no further information of the dependencies. Because the prediction interval is now held fixed, neither the changing probabilities of relapse or death nor the length of the prediction interval have any influence, and the curves drawn for each possible occurrence time of CMV-infection or chronic GvHD give exactly the same prediction; i.e., a horizontal line (which is not shown here). We comment more on this after presenting first the second application.

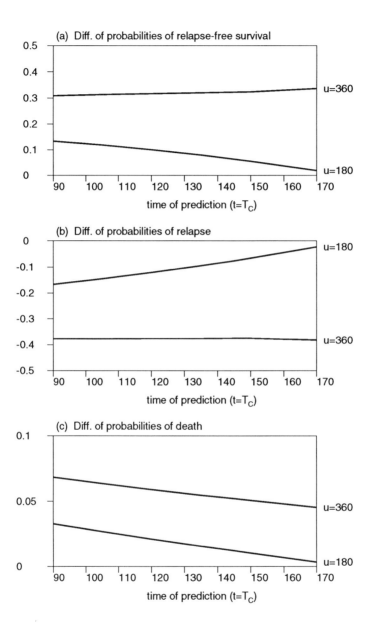

Fig. 4.8. Innovation gains, when predicting (a) relapse-free survival until u (b) relapse and (c) death in remission before u, from observing CMV-infection at t (no acute GvHD before t).

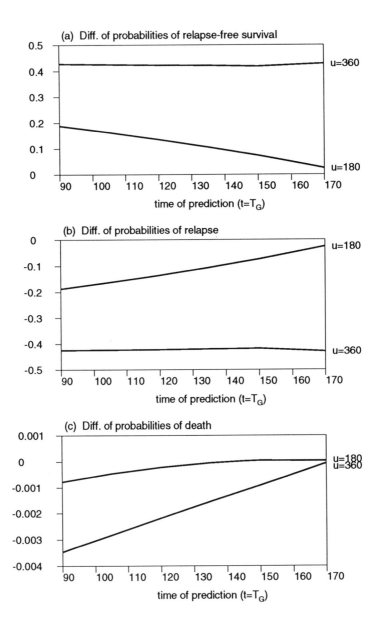

Fig. 4.9. Innovation gains, in predicting (a) relapse-free survival until u (b) relapse and (c) death in remission before u, from observing chronic GvHD at t. (no acute GvHD before t).

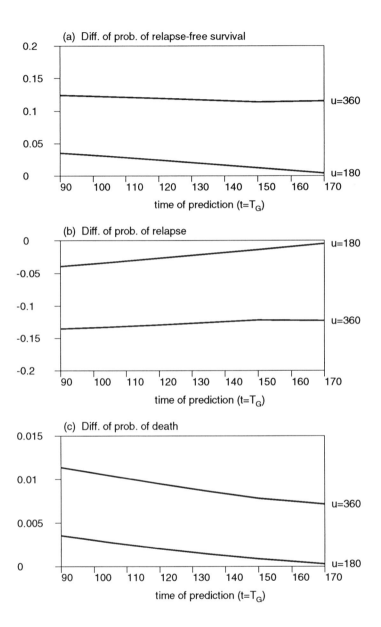

Fig. 4.10. Innovation gains, in predicting (a) relapse-free survival until u (b) relapse and (c) death in remission before u, from observing chronic GvHD at t when CMV-infection has occurred previously (no acute GvHD before t).

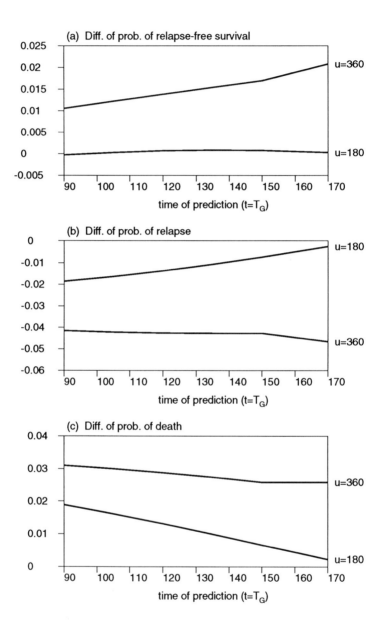

Fig. 4.11. Innovation gains, in predicting (a) relapse-free survival until u (b) relapse and (c) death in remission before u from observing chronic GvHD at t, when CMV-infection has occurred previously (acute GvHD before t).

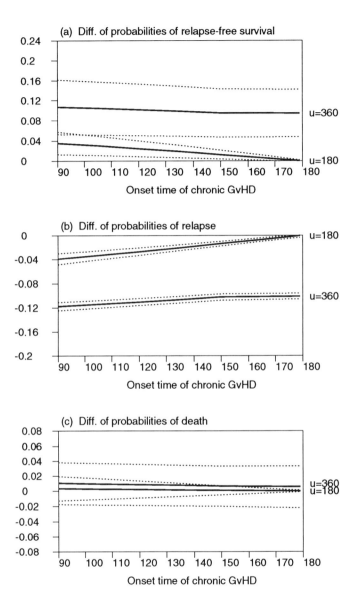

Fig. 4.12. Approximate 95%-confidence intervals for the innovation gains in predicting (a) relapse-free survival, (b) relapse and (c) death in remission from observing chronic GvHD at t, when CMV-infection has occurred previously (no acute GvHD before t).

4.3.2 The 55-Helsinki cohort: the effect of childhood separation on subsequent psychiatric admissions

In the second example the effect of childhood separation experience on later mental health is examined prospectively in a birth cohort. Several retrospective studies among psychiatric patients have revealed that an experience of separation or loss of one or both of the biological parents in childhood is common in their life-histories (e.g. Tennant et al. 1981, Väisänen, 1975). Many developmental theories also emphasize the significance of a stable relationship with the child's primary object relations. Disturbances and discontinuities in these relations are regarded harmful in many stages of the child's development (e.g. Wolfenstein, 1966, Bowlby, 1969). A more pragmatic hypothesis is that the longer a child lives in conditions which ultimately lead to the intervention of child welfare authorities, the more harmful it is. Perhaps this dual interpretation is the reason why empirical results of, for example, the importance separation age are, more or less, controversial.

In this cohort, the effect of childhood separation on subsequent mental hospitalizations was studied previously by Eerola (1989a and b) using ordinary hazard methods for recurrent events. Childhood separation was measured by registered temporary or permanent placements in children's homes or foster homes, and the placement histories were collected from the registers of child welfare authorities. It was found that in boys the effect of separation on psychiatric admissions in adolescence and in early adulthood was even strengthened when previous psychiatric care history, which otherwise was the major explanatory factor, was accounted for. This suggested that in boys (several) separations in childhood remained an important risk for severe mental health problems also in later life. For girls, the effect of separation disappeared when the psychiatric history was accounted for, whereas another indicator of instability in childhood, divorce or parental death in the family before age 7, persisted as a risk factor for subsequent psychiatric admissions in adolescence and in early adulthood. A more detailed analysis of the placement histories showed that boys were more often placed solely in institutions, whereas girls were more often placed in family-like conditions. This could partly explain the differential response among boys and girls.

In that study the age of (first) placement was measured in a rather crude way, but even then it was evident that late (first) placements increased the risk of psychiatric incidence much more than early ones. We continue here by studying the effect of age at separation explicitly by the prediction method. Since we merely want to illustrate the method we restrict the analysis to boys in this work.

4.3.2.1 The causal chain

The setting is as follows: we postulate that separation in childhood increases the risk of psychiatric admissions in later life, and that the effect is stronger if the child, after separation, shows symptoms of disorder already in childhood. Thus the probability of a psychiatric admission should in that case be higher than without any psychiatric incidence in childhood. The causal chain considered is therefore

and we denote the corresponding occurrence times by T_S, T_C and T_H.

An indication of symptoms of disorders in childhood, we used the records of outpatient psychiatric care in ages 0-15. In most cases it meant a contact in a child guidance clinic. In some (very rare) cases psychiatric incidence was a hospitalization already in childhood, but in most cases the response, hospitalization, occurred in adolescence or in early adulthood (in ages 16-27), whereas approximately half of the first placements occurred during the first year of life (see Figs. 4.13-4.14).

The ordered chain assumption $(S_{r-1} < S_r)$, introduced in Ch. 2, deserves some concern here. Even though the order of events in the chain reflects our hypothesis about the effect of separation on mental health, there is of course no logical necessity for this order in the data. It is quite possible that a first placement comes after a psychiatric incidence. This was, however, very uncommon in the data, not only because half of the first placements occurred during the first year of life. If the pattern of events in the data did not correspond to the order of our hypothesis, and we restricted only to those cases for which the events are in the "correct" order, we would undoubtedly make a serious mistake, and lose the ideal observational plan of a birth cohort. The data support our hypothesis, except for those 12 boys out of 76, for whom $T_S > T_C$. In these cases we let $T_C = \infty$.

Since the measures of child welfare authorities concern children under 16 years, the range of T_S is $[0, 15]$ (and $\{\infty\}$). Similarly, the occurrence time of child psychiatric incidence T_C is restricted to $[0, 15]$ (and $\{\infty\}$). In effect, this means that the hazards $p_S(t), p_C(t)$ and $p_{C|S}(t)$ are zero when $t \geq 16$. The first event in the causal chain is therefore the first placement, the intervening event the first outpatient contact in childhood, and the response event the first psychiatric admission in ages $[0, 28]$.

4.3.2.2 The data

The cohort consists of all children born in 1955 in Helsinki. They are representative to an age class, but of course areally and historically constrained. The cohort members were followed from birth until the end of 1982 when they reached the age of 28. From those originally 6789 (M=3506, F=3283) children the infant mortality (307 children) cases were excluded. Otherwise mortality, which was about 1% during the end of the follow-up, was not taken into account as censoring. Emigration and immigration, which could be more influential sources of bias, could not be followed during the entire period and were also ignored. Thus the only type

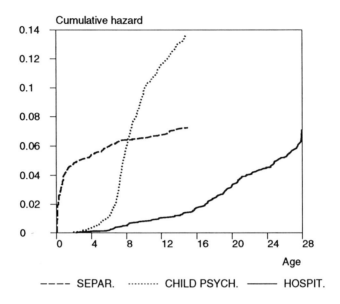

Fig. 4.13. Estimated cumulative hazards of separation, child psychiatric incidence and psychiatric admission in the 55-Helsinki cohort.

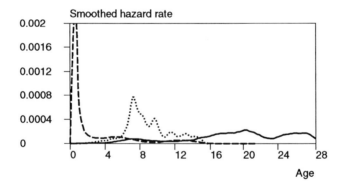

Fig. 4.14. Smoothed hazard rate functions of separation, child psychiatric incidence and psychiatric admission in the 55-Helsinki cohort.

of censoring is right-censoring at the end of the follow-up 31.12.1982. The follow-up data were collected from the national discharge register for mental hospitals, patient registers of psychiatric clinics and mental health bureaus throughout the country, and from registers of child welfare authorities. A more comprehensive representation of the material can be found in Amnell (1974) and Almqvist (1983).

Since children diagnosed as mentally retarded or subnormal were always involved with a psychiatric contact in childhood, and some of them were also taken into custody because of this defect, all those children (F=81, M=150) were excluded. This makes the sizes of the separation and childhood incidence groups much smaller but prevents from trivial conclusions. We show the number of cases for both boys and girls but, as mentioned already, we perform the analyses only for boys.

Table 4.3 Absolute numbers of psychiatric admission, placements and childhood psychiatric incidence.

	Psychiatric admission				
	no Separation		yes Separation		
	no	yes	no	yes	Total
WOMEN:					
no childpsych. incidence	2533	135	110	11	2789
childpsych. incidence	175	42	11	8	236
Total	2708	177	121	19	3075
MEN:					
no childpsych. incidence	2488	135	133	11	2767
childpsych. incidence	309	60	24	16	409
Total	2797	195	157	27	3176

4.3.2.3 The hazard models

Since the focus of the analysis was on the influence of the ages at separation and child psychiatric incidence on the prediction of a psychiatric admission, and the number of admissions is quite small, we used fairly simple models for the causal events. More detailed information about each event was taken into account in Eerola (1989a and b). So, we do not claim that these models are in any sense complete descriptions of the effect of separation on later psychiatric health. The figures in Table 4.3 do not support the hypothesis "Separation causes *the* psychiatric admissions in adolescence and in early adulthood" since 14% of those who experienced separation and 6% of those who did not, have subsequent admissions.

The application serves merely as an example of a setting in which one could not even think of being able to measure all possible influencing factors.

In a preliminary analysis we investigated the influence of occurrence times of S and C by stratified Nelson-Aalen plots. Separation age was classified as: < 1, 1-2, 3-5, 6-15 years, and age of child psychiatric incidence as: 0-5, 6-8, 9-11, 12-15 years. The stratified cumulative hazards were calculated both on the real time scale (on T_H) and on the relative time scale ($T_H - T_S$ or $T_H - T_C$) where the origin is determined by the occurrence times T_S and T_C. The use of these two time scales corresponds to different hypotheses about the causal effects (cf. Section 4.2), and we return to that in Section 4.6.

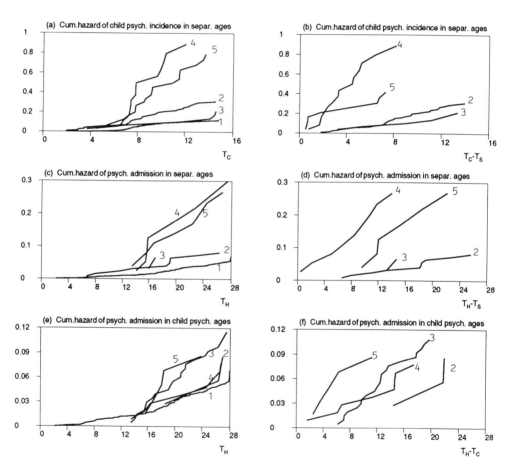

Fig. 4.15. Stratified Nelson-Aalen plots of (a) childhood incidence in separation age strata: 1=no separation, 2=before age 1, 3=ages 1-2, 4=ages 3-5, 5=ages 6-15 on the scale T_C, (b) on the scale $T_C - T_S$, of (c) psychiatric admission in separation age strata on the scale T_H, (d) on the scale $T_H - T_S$, of (e) psychiatric admission in childhood incidence age strata: 1=no incidence, 2=ages 0-5, 3=ages 6-8, 4=ages 9-11, 5=ages 12-15 on the scale T_H and (f) on the scale $T_H - T_S$.

In Fig. 4.15 (a)-(f) the estimated cumulative hazards of child psychiatric incidence are shown in four separation age groups. Although the influence of age at separation is not completely monotonic, it is clear that late separations increase the risk of childhood incidence more than early ones. The number of subsequent psychiatric admissions is particularly small in group 3 (1-2 years). The same applies to the cumulative hazards of psychiatric admission (Fig. 4.15 (c)-(d)) in the four separation age strata. The difference between groups 4 and 5 is rather small when considering the hazards on the time scale T_H (Fig. 4.15 (c)), so we modelled the effect of T_S instead of that of the interval $T_H - T_S$ in the hazard models of H and C.

The influence of T_C on the risk of admission is small and rather unclear on the time scale T_H (Fig. 4.15 (e)), but when plotting the cumulative hazards of H on the scale $T_H - T_C$ (Fig. 4.15 (f)), it is clear that the earlier C occurs (compared to H) the higher is the risk of H, although the difference between groups 2,4 and 5 is very small. We therefore modelled the effect of $T_H - T_C$ instead of that of T_C in the hazard model of H. The results of the modelling are in Appendix 2. In estimation we used the grid value of 30 days which is a good approximation of continuous time scale even though the events were recorded at exact dates. The possible ties in the data do not, of course, concern us in the discrete time model.

Separation. Since half of the first placements occurred during the first year of life, possible explanatory covariates must relate to that time. Two fixed covariates were used: the *social status of the family in 1955* (0=upper and 1=lower) and *marital status of the mother at the time of birth* (0=married, 1= unmarried, divorced or widowed). The model is

$$ln(\frac{P(\Delta N_t(S) = 1|\mathcal{F}_{t-1})}{P(\Delta N_t(S) = 0|\mathcal{F}_{t-1})}) = \alpha_0 + \alpha_1 1_{\{SOC55=1\}} + \alpha_2 1_{\{MARIT55=1\}} + \alpha_3 log(t)$$

Child psychiatric incidence. Two fixed covariates, *social status in 1955* and *stability of the childhood family* (1= divorce or parental death before age 7, 0=neither) were both significant risk factors for childhood incidence. Although the second covariate was available only as a fixed indicator, the number of hospitalizations before age 7 is so small that it causes no serious bias. Several transformations of the age at separation were tried but we show only the results of the best model in which the effect of separation was modelled as $1_{\{T_S \leq t\}}(\beta_1 + \beta_2 T_S/30)$. The parameter β_1 measures the indicator effect of separation (the level from which the conditional hazard $p_{C|S}(t|v)$ starts at T_S) regardless of T_S, and β_2 measures the additional effect of separation *age* in months.

As expected from earlier studies, separation increases the risk of child psychiatric incidence ($\hat{\beta}_1 = 0.48$, s.e.=0.09) and the effect of age at first placement is also highly significant so that late placements increase the risk of childhood incidence more than early placements ($\hat{\beta}_2 = 0.01, s.e. = 0.003$). The model is

$$ln(\frac{P(\Delta N_t(C) = 1|\mathcal{F}_{t-1})}{P(\Delta N_t(C) = 0|\mathcal{F}_{t-1})}) = \beta_0 + \beta_1 1_{\{SOC55=1\}} + \beta_2 1_{\{FAM69=1\}}$$
$$+ \beta_3 1_{\{7 \leq t < 12\}} + \beta_4 1_{\{12 \leq t < 16\}} + \beta_5 1_{\{0 \leq t < 7\}} \cdot t$$
$$+ \beta_6 1_{\{7 \leq t < 12\}} \cdot t + \beta_7 1_{\{12 \leq t < 16\}} \cdot t$$
$$+ \beta_8 1_{\{T_S \leq t\}} + \beta_9 1_{\{T_S \leq t\}} T_S/30$$

Psychiatric admission. The same transformation of T_S as above, $1_{\{T_S \leq t\}}(\gamma_1 + \gamma_2 T_S/30)$, gave the best fit also when modelling the effect of separation on the hazard of psychiatric admission. Late placements increased the risk of admission significantly more than early ones ($\hat{\gamma}_2 = 0.01, s.e. = 0.003$). However, separation as such (the indicator) lost significance ($\hat{\gamma}_1 = 0.13, s.e. = 0.30$) when the term of separation age was included. We show the results of the best fitting model in which a transient effect of C is assumed; i.e., $1_{\{T_C \leq t\}}(\gamma_1 + \gamma_2 exp(-\gamma_3(t - T_C)/365))$. However, it turned out that the effect of C actually did not disappear. In a transient effect model the parameter of $t - T_C$ in the exponent is negative, and since we estimated the model with that (negative) γ_3 which gives the best fit ($\hat{\gamma}_3 = -0.05$), the coefficient of $exp(-0.05(t - T_C)/365)$ became in turn negative ($\hat{\gamma}_2 = -1.49$, s.e.=1.09) which actually results in a slowly increasing curve as a function of $t - T_C$. We return to the different transformations in Section 4.4. The model is

$$ln(\frac{P(\Delta N_t(H) = 1|\mathcal{F}_{t-1})}{P(\Delta N_t(H) = 0|\mathcal{F}_{t-1})}) = \gamma_0 + \gamma_1 1_{\{SOC55=1\}} + \gamma_2 1_{\{FAM69=1\}}$$
$$+ \gamma_3 1_{\{6 \leq t < 15\}} + \gamma_4 1_{\{15 \leq t < 21\}} + \gamma_5 1_{\{21 \leq t < 28\}}$$
$$+ \gamma_6 1_{\{T_S \leq t\}} + \gamma_7 1_{\{T_C \leq t\}} + \gamma_8 1_{\{T_S \leq t\}} T_S/30$$
$$+ \gamma_9 1_{\{T_C \leq t\}} exp(-0.05(t - T_C)/365)$$

4.3.2.4 The prediction probabilities

The formulas of the prediction probabilities in the 3-point chain are the same as those in Chapter 3. The derivatives for constructing their confidence limits are in Appendix 1.

4.3.2.5 Fixed starting point t, fixed history H, endpoint u varies

We consider again first the situation in which the endpoint u is let to vary, now in the interval $u \in (t, 28]$. The hypothetical individual, for whom the predictions are shown, comes from lower social status family (in 1955) and has experienced divorce (or death) of the parents before age 7. We show in Fig. 4.16 the probability of $\{T_H > u\}$ at three particular values of the time of prediction, $t = 3, 7$ and

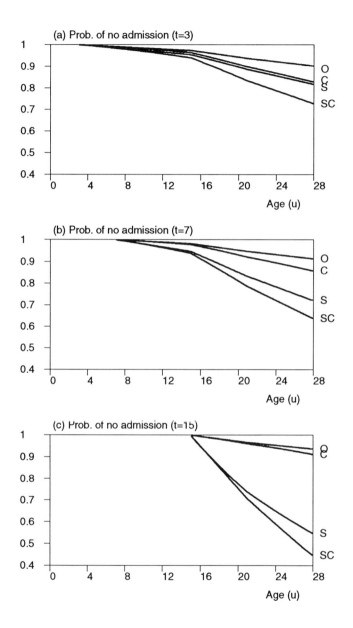

Fig. 4.16. Probabilities of no psychiatric admission before u for alternative histories: S=separation, C=child psyhiatric incidence, SC=both, O=neither at t; time of prediction $t=T_S$ or T_C (a) $t=3$ (b) $t=7$ (c) $t=15$ ($u=28$).

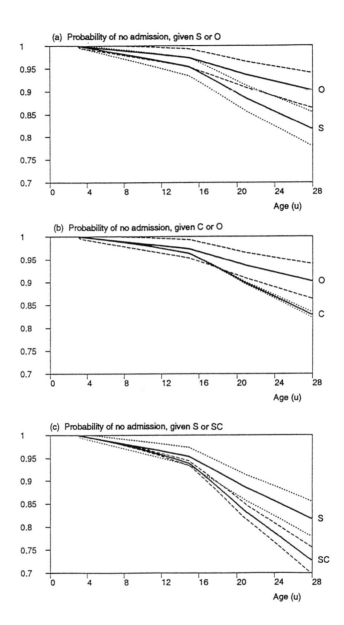

Fig. 4.17. Approximate 95%-confidence intervals for the probability of no admission before age 28 when comparing histories with (a) $T_S = t = 3$ vs. $T_S > t, T_C > t$ (b) $T_S = \infty$, $T_C = t = 3$ vs. $T_S > t, T_C > t$ (c) $T_S = 1$, $T_C = t = 3$. vs. $T_S = 3$, $T_C > t = 3$.

15, corresponding to the occurrence times of either T_S or T_C, to demonstrate how different answers we can get depending on the time of prediction. Even though the probabilities of $\{T_H > u\}$, given the possible histories $\{(v, S)\}$, $\{(w, C)\}$, $\{(v, S), (w, C)\}$, $v \leq w \leq t$, and $\emptyset$, are in the same order at $u = 28$ at all three t-values, the influence of separation or child psychiatric incidence depends clearly on their occurrence time. At age 3 they are equally hazardous, whereas child psychiatric incidence as late as at age 15 has practically no influence on the probability of psychiatric admission before age 28. The fact that the prediction is worst when both S and C have occurred by t, supports our initial hypothesis.

Instead of confidence limits for the innovation gains, in which case we compare the prediction and its confidence limits at a particular u-value and vary the time of prediction t, we compare in this case the entire no admission curves, and their 95%-confidence limits, at a particular t-value, $t = 3$. The confidence limits of the curves with histories $H_3 = \emptyset$ and $H_3 = \{(3, S)\}$ (Fig. 4.17 (a)) overlap in almost every point in the prediction interval $(3, 28]$, suggesting that when separation occurs as early as at $T_S = 3$, conclusions about its effect on the admission probability are rather uncertain. Because the effect of separation increased with occurrence age, a later time of prediction $t = T_S$ would probably result in larger differences between the curves. The confidence limits of the probability of $\{T_H > u\}$ with the history $H_3 = \{(3, C)\}$ (Fig. 4.17 (b)) are extremely narrow mainly because the only randomness is in the occurrence of the response, but also because the sample size of those who experienced C is quite large ($n_C = 409$). In contrast, the limits of the curve with the history $H_3 = \{(1, S), (3, C)\}$ in Fig. 4.17 (c) are wider because the sample size in only $n_{SC} = 76$. The curves in Fig. 4.17 (c) are not directly comparable since in the S-curve $T_S = 3$ whereas in the SC-curve $T_S = 1$ and $T_C = 3$. To sum up, given the additional information about the confidence limits, the effects of both separation and child psychiatric incidence are small and rather uncertain when their occurrence time is as early as at age 3.

4.3.2.6 Fixed history H, fixed endpoint u, starting point t varies

Instead of presenting a separate curve for each t we again calculate the cross-sections of the no admission probability at prespecified values of u ($u = 21$ and 28). The time of prediction t varies now in the interval $[0, 15]$, which is the range of possible values for T_S and T_C. We compare predictions of the form $t \mapsto \mu_t^*(H \cup \{(t, x)\}; I)$, where H is either $\{\emptyset\}$ or $\{(v, S)\}$ with $v < t$, and $x = S$ or C depending on whether S has already occurred or not, and $I = (3, u]$.

In Figures 4.18 (a) and 4.18 (b) we compare the probabilities of no admission before $u = 21$ or 28 when either S or C has just occurred vs. has not yet occurred (O). The risk of admission increases clearly by separation age. There is about 50% chance for an individual experiencing the first placement as late as at age 15 to be hospitalized before age 28. The age of child psychiatric incidence has an opposite effect; the earlier it occurs the higher is the risk for hospitalization before age 28, provided, of course, that the model gives a reasonable description of the data. Since the slopes are rather small, the decreasing curve of T_C can be a consequence

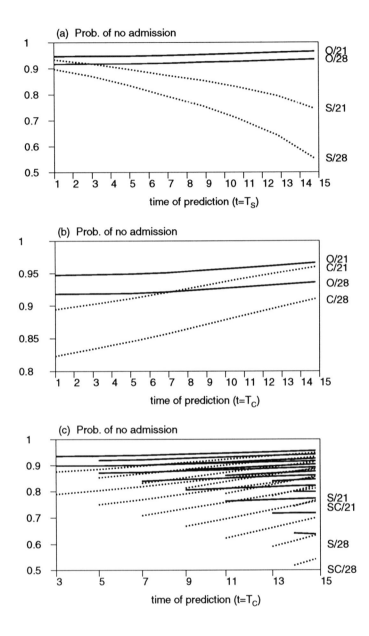

Fig. 4.18. Probabilities of no psychiatric admission before ages 21 and 28 when (a) separation (b) child psychiatric incidence (c) child psychiatric incidence, given separation previously, has just occurred (S,C or SC) or has not yet occurred (O or S). In (c) each curve corresponds to a particular value of $T_S = 1, 3, .., 13$. (S=separation, C=child psyhiatric incidence, SC=both, O=neither)

of shortening prediction interval only. We shall check that in 4.3.2.7 below. Note that in Figures 4.18 (a) and 4.18 (c) the curves denoted by $S/21$ or $S/28$ are not the same; in (c) is assumed that $T_S = v < t$ and $T_C > t$, whereas in (a) $T_S = t$ and $T_C > t$.

The innovation gains from child psychiatric incidence, given that separation occurred previously (Fig 4.19 (c)), are larger than without preceding separation, and they are the larger the later separation occurred until age 9. Thereafter, when the hazard of H increases, and on the other hand the possible interval between H and C becomes smaller (as $t \to 15$), the effect of C decreases. Since the innovation gains from C are larger when S has occurred previously, we investigated the possibility that those who experience C after S have a different overall risk for hospitalization than those who did not. We considered a model with an additional time-dependent interaction term $1_{\{T_S \leq t\}} \cdot 1_{\{T_C \leq t\}}$ between S and C. The coefficient of this term was positive but not significant ($-2logL = 0.76$, $df = 1$). However, the parameter of the indicator $1_{\{T_S \leq t\}}$ became in this model negligible, indicating that without a subsequent child psychiatric incidence separation alone did not increase significantly the risk of admission, unless it occurs late since the coefficient of $T/30$ remained highly significant also in this model. We calculated the innovation gains from C (given S any time previously) also in the interaction model (not shown here), and they appeared to be about 5% larger than those in the model without the interaction term.

4.3.2.7 Fixed endpoint u, fixed starting point t, history H varies

Finally, we consider the effects of S and C on H when the influence of the length of the prediction interval is eliminated. The net influence of T_S (Fig. 4.20 (a)) is clearly increasing (i.e.; the probability of $\{T_H > u\}$ is decreasing with T_S) whereas the net influence of T_C is much more moderate than was shown for the innovation gains (cf. Fig. 4.19 (b)). The differences in Figs. 4.19 (c) and 4.20 (c) show the effect of the length of the prediction interval. In Fig. 4.19 (c) the distance between the prediction curves at different T_S-values becomes smaller the shorter the prediction interval is (i.e., as $t \to 15$), whereas in Fig. 4.20 (c) the curves are almost horizontal, indicating a very weak influence of the occurrence time of C on the probability of H. Note that the later S occurs the larger are the differences between the curves. This effect disappears in the curves of the innovation gains because in both probabilities (i.e., in $P(T_C > u|SC)$ and in $P(T_C > u|S)$) S is assumed to have occurred at the same time v.

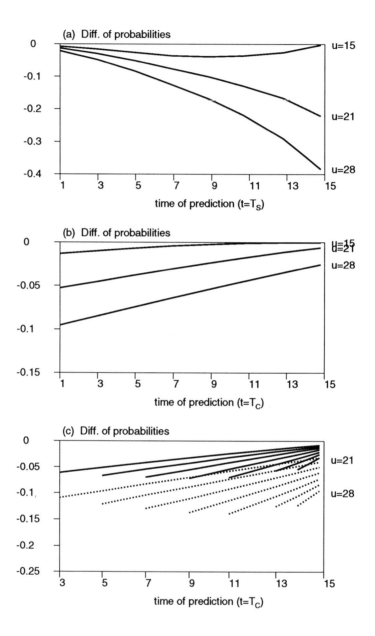

Fig. 4.19. Innovation gains in predicting psychiatric admission before ages 21 and 28, from observing (a) separation (b) child psychiatric incidence (c) child psychiatric incidence when separation has occurred previously. In (c) each curve corresponds to a particular value of $T_S = 1, 3, .., 13$. (S=separation, C=child psyhiatric incidence, SC=both, O=neither)

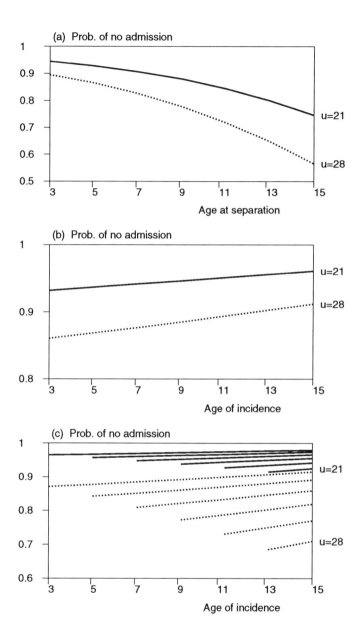

Fig. 4.20. Dependence of the probability of no admission before ages 21 and 28 on the occurrence time of (a) separation (b) child psychiatric incidence and (c) child psychiatric incidence when separation has occurred any time previously. Time of prediction=15.

4.4 Sensitivity of the innovation gains on hazard specification

In Chapter 1 our argument against using regression coefficients as basic causal concepts was that they are entirely dependent on a particular specification of a regression model. Changing the model specification changes the estimated "amount of causal dependence". Our suggestion was to use the innovation gains from observing the occurrence of a cause when predicting the event of interest. It is expected that the prediction probabilities are less sensitive to modelling because they are (parametric forms of) cumulative distribution functions (or their complements) which in the uncensored case correspond to the empirical distribution function; i.e., the distribution of the frequencies in the data. However, the prediction probabilities are also functions of the hazards, which themselves are estimated by regression models, and are thereby prone to be affected, at least to some extent, by modelling. In this section we study the sensitivity of the innovation gains on different specifications of the hazards.

At least three possible sources of sensitivity can be mentioned: the type of the hazard models used, the specification of the basic time variable, and the functional form used to model time-dependence between the events. Usually different types of hazard models (e.g. Cox's model, piecewise exponential model) give fairly similar estimates for the regression coefficients, and since we used small grid values in estimation, we are not concerned about this source of sensitivity. In a fully parametric model the basic underlying time is parameterized also, and it can usually be done in a rough way if the main interest is in the regression coefficients. If the hazard rate varies very much within the prediction interval, "wrong" specification of the time variable, for example, by piecewise constant intervals could in principle affect the prediction probability at a particular time. However, it is expected that such a "wrong" specification of time variation does not affect the estimated innovation gains very much because the underlying time is modelled in the same way in both probability terms (e.g. $P(T_C > u|A)$ and $P(T_C > u|\emptyset)$) considered.

When considering the innovation gains, it is expected that a more influential source of sensitivity is the functional form used to model time-dependence between the causal events, especially if these effects are strong. As we saw in the two examples, it is not always easy to decide what the "right" functional form of dependence should be, and in many practical cases there exists no a priori hypothesis either. In such situations it is important to know how robust the prediction probabilities are on different functional specifications. Obviously this is partly a question of goodness-of-fit, but it is also well-known that a data set can support two entirely different hypotheses. The question of modelling either S_r or $S_r - S_{r-k}$, $r > k \geq 1$, is related to the importance of absolute (real or calendar time) vs. relative (follow-up time) interpretation of time. The hypothesis "Early (vs. late) occurrence of a cause increases the probability of the effect" relates to the actual occurrence time S_r, whereas a hypothesis such as "The effect of a cause is transient (vs. cumulative)" relates to the time between a cause and its response and not to the actual occurrence time. In the first case the interpretation of time as real time is important and the time origin cannot be changed, whereas in the second case a new time origin is defined at each S_r.

We consider next a set of hazard models in which different transformations were tested. Some examples of possible functional forms were presented in Section 4.2, and several of these (not all shown here) were tested for the 55-cohort data - the number of cases in the BMT data was too small for this purpose. We restricted to logistic hazard regression models with exponential risk function so that the effect of each covariate on the logit-hazard is additive in the exponent. In all models we make a distinction between the "occurrence time independent" effect of e.g. S, which is $1_{\{T_S \leq t\}}$, and the additional, on the "occurrence time dependent" effect $1_{\{T_S \leq t\}} \cdot g(T_S)$. To simplify Table 4.4, we present $g(\cdot)$ without the indicator but of course all these terms are 0 when $t < T_S$. We refer to Section 4.2 for the interpretation of each transformation.

Table 4.4 *Different transformations of T_S and T_C in the models for the hazard of psychiatric admission (the 55-cohort)*

Model	Transformation	$-2\log L$
1	$1_{\{T_S \leq t\}} + 1_{\{T_C \leq t\}}$	3424.53
2	$1_{\{T_S \leq t\}} + log(T_S) + 1_{\{T_C \leq t\}} + log(T_C)$	3416.86
3	$1_{\{T_S \leq t\}} + log(t - T_S) + 1_{\{T_C \leq t\}} + log(t - T_C)$	3418.85
4	$1_{\{T_S \leq t\}} + T_S/30 + 1_{\{T_C \leq t\}} + T_C/30$	3413.55
5	$1_{\{T_S \leq t\}} + T_S/30 + 1_{\{T_C \leq t\}} + log(t - T_C)$	3411.41
6	$1_{\{T_S \leq t\}} + T_S/30 + 1_{\{T_C \leq t\}} + exp(-0.05(t - T_C)/365)$	3411.65

In all six models the changes in the log-likelihoods due to $g(T_S)$ and $g(T_C)$ were significant compared to the simple constant effect model (Model 1). Models 4, 5 and 6 are practically equally good in the likelihood sense but more close inspection of the single coefficients (Table 4.5) shows that the transformations used in model 5, $g(T_S) = T_S/30$ for S and $g(T_C) = log(t - T_C)$ or alternatively $exp(\gamma(t - T_C))$, $\gamma = -0.05$ in model 6 for C, give the best fit.

The value $\gamma = -0.05$ in model 6 was obtained by maximizing the likelihood at certain fixed values of γ with respect to all other parameters (i.e., by specifying the profile likelihood of γ at those values) and choosing that value of γ which gives the largest value for the log-likelihood.

The transformation $g(T_C) = log(t - T_C)$ is problematic if $\beta < 0$ because when $t \to T_C$, we have $(t - T_C)^\beta \to \infty$ when considering an exponential risk function. The range of possible values of β are therefore restricted. In the data, however, $\hat{\beta} > 0$.

When inspecting the regression coefficients in Table 4.5 it is obvious that different transformations yield quite different coefficients for both $1_{\{\cdot\}}$ and $g(\cdot)$. Not only the size but also the sign of the regression coefficients vary considerably even though the differences in the log-likelihoods are small. Without the $g(\cdot)$-terms the indicators of both S and C are very significant (Model 1). However, except for Model 3, in which the transformation $g(T_S) = log(t - T_S)$ is clearly unsuitable for

T_S, in all other models the indicators become insignificant when the terms $g(T_S)$ and $g(T_C)$ are included.

Table 4.5. Estimated regression coefficients of S and C in Models 1-6

Term	Model 1	Model 2	Model 3	Model 4	Model 5	Model 6
$1_{\{T_S \leq t\}}$	0.6382	-1.4392	3.9382	0.1253	0.1292	0.1278
s.e.	(0.2199)	(0.8648)	(1.4779)	(0.2979)	(0.2974)	(0.2974)
$1_{\{T_C \leq t\}}$	0.6123	0.6702	-2.5812	0.6388	-2.3067	1.4433
s.e.	(0.1850)	(3.6506)	(2.3149)	(0.5178)	(2.1291)	(0.6708)
$g(T_S)$	-	0.3394	-0.3921	0.0113	0.0117	0.0112
s.e.	-	(0.1272)	(0.1766)	(0.0031)	(0.0031)	(0.0031)
$g(T_C)$	-	-0.0132	0.3943	-0.0008	0.3522	-1.4911
s.e.	-	(0.4538)	(0.2602)	(0.0044)	(0.2596)	(1.0928)

However, we still kept the indicators in the model to be able to separate the additional effect of the occurrence time. The effect of $g(T_C)$ is nearly significant in Models 5 and 6, but on the whole very small compared to $g(T_S)$. In fact, for C it would be enough to know that it has happened by time t; i.e., to model its occurrence just as an indicator.

4.4.1 The innovation gains from observing S and C in Models 1-6

As expected, the innovation gains are much less sensitive to transformations than the regression coefficients. We show first the cross-sections of the prediction probabilities for each model at times $u = 21$ and $u = 28$.

To study the effect of both S and C in different transformations we show in Figures 4.21 (a)-(f) the predictions of no admission when separation has occurred anytime previously and child psychiatric incidence has just occurred (SC), or has not yet occurred (S). Except for Models 1 and 3, the figures are strikingly similar. The main reason is, of course, that $g(T_S) = T_S/30$ in all these models. The difference of Model 3 from Models 4-6 is due to modelling the effect of S as $log(t - T_S)$. The negative coefficient of $log(t - T_S)$ could be an indication of a "latent" effect of early separations which then result in psychiatric admissions in early adulthood, but more probably it is just a consequence of the fact that most separations occur very early and that there are very few hospitalizations before age 15.

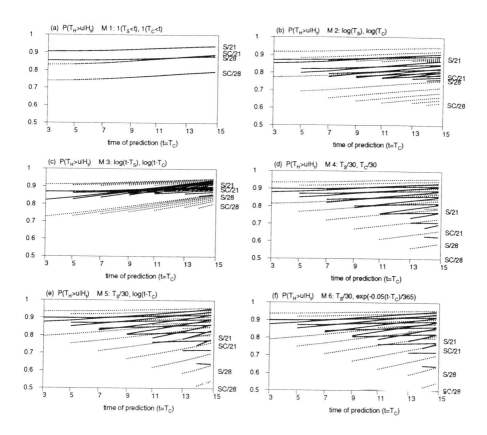

Fig. 4.21. Probabilities of no psychiatric admission before ages 21 and 28 in Models 1-6 when child psychiatric incidence, given separation previously, has just occurred (SC) or has not yet occurred (S). Each curve corresponds to a particular value of $T_S = 1, 3, .., 13$.

The effect of T_S in Model 3 (with $g(T_S) = log(t-T_S)$) is much smaller than when modelling the actual occurrence time, and this is the reason for smaller differences *between* the curves starting at $t = 3, 5, ..., 15$. The logarithmic transformation of T_S in Model 2 can, on the other hand, be seen in decreasing intervals between the curves as $T_S \to 15$. The *slope* of the curves is, however, almost the same in all figures, indicating a weak effect of T_C. The innovation gains in Figures 4.22 (a)-(f) are even more stable than the prediction curves, and vary between -0.01 and -0.15 depending on T_S and to some extent on T_C in all models. Regardless of T_S and T_C, the innovation gains from observing a child psychiatric incidence, when separation has occurred previously, are positive for the probability of admission before age 28, the larger the later the first separation occurs and the earlier an incidence occurs, although the age effect is weak. This, and the effect of a longer prediction interval, are the reasons why the innovation gains are for all t larger in the long-term predictions ($u = 28$) than in the short-term predictions ($u = 21$). The earlier the events occur, the more they increase the cumulative probability of a later admission.

4.4.2 The estimated hazards with different transformations

We continued the analysis by comparing the additional changes in the hazards due to the transformations in Models 2-6. Because the fixed covariates and basic time covariates are the same in each model, we plotted the curves $[1 + exp(-\hat{\alpha}_1 1_{\{T_S \leq t\}} - \hat{\alpha}_2 g(T_S))]^{-1}$ using different $g(T_S)$ (resp. $g(T_C)$) and their estimated coefficients.

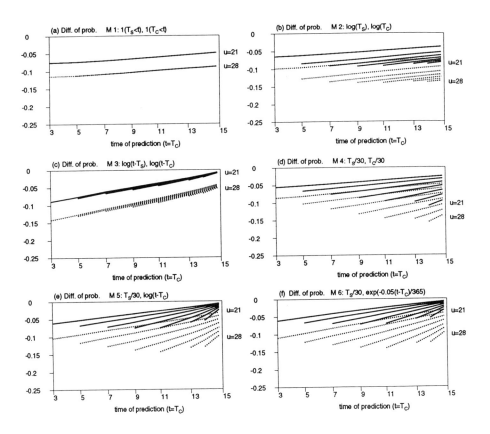

Fig. 4.22. Innovation gains, in predicting psychiatric admission before ages 21 and 28, from observing child psychiatric incidence in Models 1-6 when separation has occurred previously. Each curve corresponds to a particular value of $T_S = 1, 3, .., 13$. (S=separation, C=child psyhiatric incidence, SC=both, O=neither)

As expected, the transformations differ mostly at the extreme areas of the ranges of T_S and T_C.

The two transformations of S, $g(T_S) = T_S/30$ and $log(T_S)$, give rather deviating curves and the difference increases rapidly when $T_S > 8$ (Fig. 4.23 (a)). The logarithmic increase seems, however, too moderate for the data. As mentioned, modelling the time between H and S is rather problematic, and we do not consider the figures of these transformations here. The same two transformations of T_C ($g(T_C) = T_C/30$ and $log(T_C)$) give almost identical fits (Fig. 4.23 (b)). The weak effect of T_C can also be noted from Fig. 4.23 (c) in which completely different hypotheses about the effect of C (i.e., transient effect with $exp(-0.05(t - T_C))$ and cumulative effect with $log(t - T_C)$) give almost identical fits in the observed range of $(T_H - T_C)$. This range is marked by a vertical line in the figures.

An alternative model (MODEL 7, not shown elsewhere), where γ was positive ($\gamma = 0.15$), gave equally good fit in terms of the likelihood ($-2logL = 3411.3$), but it is easy to see that when extrapolating outside the observed range of T_C we get totally different values than from the other transformations.

Because the effect of C is so weak, it is understandable that models 4,5 and 6, in all of which $g(T_S) = T_S/30$, have log-likelihoods of almost the same size. These plots show clearly that one should be careful when modelling time-dependence between the events. Especially, if the number of cases is small, it can be quite misleading to try to capture the time variation in some simple functional form. At least different possibilities should be inspected carefully in the light of the observed range of the occurrence times.

4.4.3 Residual plots with time-dependent covariates

The values of the log-likelihoods, which are sums calculated at the end of the follow-up, give, of course, a very poor picture of the goodness-of-fit *in time*. Several authors have recently suggested the use of residuals in model-checking also in survival analysis (e.g. Schoenfeld (1982), Cain & Lange (1984), Arjas (1988), Barlow & Prentice (1988), Therneau et al. (1990)). Therneau et al. consider the functional form specification of a covariate Z^* in the Cox model, and suggest that (when ignoring Z^*) a smoothed plot of the residuals $\hat{M}_i^*(T_i) = N_i(T_i) - \Lambda_0(T_i)exp(\beta'Z_i)$, T_i being the individual observation time, against the values of Z^* would give approximately the correct form in the exponent. Their method assumes, however, that the covariates are fixed. Since time-dependent covariates are an essential part of our chain model we considered residuals which account for the random occurrences of the causal events. Such residuals are of the form

$$\Delta M_t(k) = \sum_{i=1}^{n_k} 1_{\{T_{i,k} \leq t\}} \left(\Delta N_t(i) - P(\Delta N_t(i) = 1|\mathcal{F}_{t-1})\right), \qquad (4.9)$$

where $P(\Delta N_t(i) = 1|\mathcal{F}_{t-1}) = 1/(1 + exp(-\beta'Z_{t-1}(i))$, and k indicates group S for individual i if $T_{i,S} \leq t$, and group C if $(T_{i,S} \leq) T_{i,C} \leq t$, otherwise $k = \emptyset$.

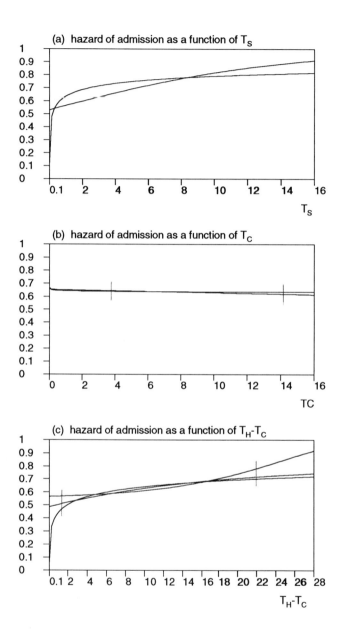

Fig. 4.23. Estimated hazards $exp(\cdot)/(1 + exp(\cdot))$ using the indicators and transformations $g(\cdot)$ (a) of T_S in models 2 and 4, (b) of T_C in models 2 and 4 and (c) of $(t - T_C)$ in models 3,6 and 7 (see Table 4.4). Vertical lines denote the observed ranges of T_C.

An individual experiencing S at T_S and C at T_C thus contributes to group $\emptyset$ at $t < T_{i,S}$, group S at $T_{i,S} \leq t < T_{i,C}$, and group C at $t \geq T_{i,C}$.

For the individual residuals $M_t(i)$ holds that $E(M_t(i) - M_{t-1}(i)|\mathcal{F}_{t-1}) = 0$, $i = 1,..,n$, and therefore the processes which are sums of the differences $\Delta M_t(k)$ over time

$$M_t(k) = \sum_{s \leq t} \sum_{i=1}^{n_k} 1_{\{T_{i,k} \leq t\}} (\Delta N_s(i) - P(\Delta N_s(i) = 1|\mathcal{F}_{s-1})), \quad k = S, C \quad (4.10)$$

are also locally bounded martingales which should have no trend.

The joint curves of $M_t(S)$, $M_t(S)$ and $M_t(\emptyset)$ together with $M_t(T) = M_t(S) + M_t(C) + M_t(\emptyset)$ are shown in Figs 4.24 (a)-(d). We consider first the sequences of residuals of separation (S) and child psychiatric incidence (C) groups, the last of which includes also those who experienced separation before T_C. There are only 11 cases in the S-group, and when $t > 14$ all models tend to overestimate the number of hospitalizations. This overestimation could be reduced by including the time-dependent interaction term $1_{\{T_S \leq t\}} \times 1_{\{T_C \leq t\}}$, because in this model the indicator $1_{\{T_S \leq t\}}$ became negligible and thus reduced the model-given counts. The coefficients of the terms concerning C remained practically unchanged, so this did not improve the fit in other groups. Model 3, in which $g(T_S) = log(t - T_S)$, gives the poorest fit whereas models 2,4,5 and 6, which all use the transformation $g(T_S) = T_S/30$, give practically the same result. The indicator model (Model 1) overestimates the number of cases throughout.

The residuals in group C form two distinctive groups. Models 2,3,5 and 6 give identical fits since the effect C is very weak. They underestimate the number of cases slightly in ages 16-22 but models 1 and 4 (the indicator model and $T/30$ for both S and C) overestimate the number of cases in all ages. Different transformations caused small changes also in the values of the fixed covariates and the underlying time covariates which is the reason for variations in the $\emptyset$-group. These changes are balanced in the total which is the sum of all group residuals. Except for minor changes the fit is equally good in all models. The overall fit is clearly dependent on many other factors than those measured in the hazard models, and it cannot be changed very much by the time transformations. More dense division of the underlying time intervals would change the pattern of the residuals in the figure of total residuals but because the main interest was in the innovation gains, we did not elaborate on that.

4.5. Discussion

In the previous section we illustrated the uses and limitations of the prediction method with two rather different data sets. An ideal data set for illustration purposes would of course be a set of complete event-histories, enough cases in all groups of events, well-defined hypotheses about the dependence structure, and no artificial restrictions on the occurrence times of the events. In hypothetical examples, like the one presented in Chapter 2, this is possible. Since we, however,

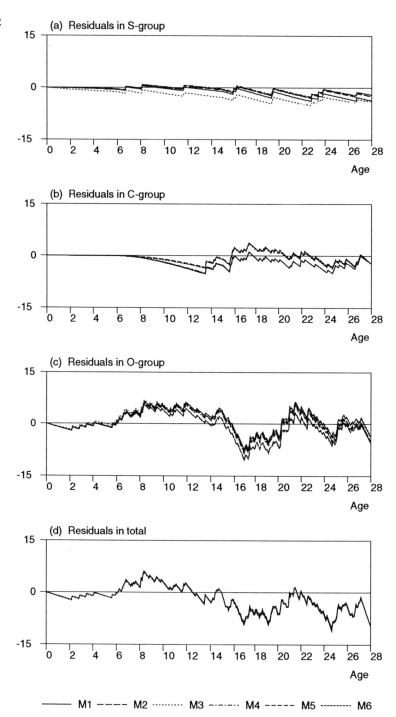

Fig. 4.24. Sums of the martingale residuals from models 1-6 of individuals for whom (a) separation (b) child psychiatric incidence or (c) neither, has occurred before t, and (d) the sum of (a)-(c).

wanted to test the method in real applications, we next evaluate its usefulness in the two examples.

The protective effect of chronic GvHD on leukemia relapse has been established in many studies, and even though the biological processes behind the dependencies are not thoroughly known yet, there is fairly good understanding among clinicians about the factors which affect relapse or death. Because of the complexity of the interacting causes it seems that a chain model, which directly models the process of recovery after transplantation, is much more informative than separate hazards of each event. The main limitation (as regards the prediction method) of the otherwise exceptionally rich data set is the small number of cases which hindered from estimating complicated time-dependencies. This is, of course, typical in clinical studies. However, even then the innovation gains provide a means to study the effect of a causal event in different perspectives, and helps to answer questions such as: "How much does the prediction of a patient with certain characteristics change within the next x months if a particular complication C occurs at time t, given what has happened to the patient so far?" We are completely free to vary either the initial characteristics of the patient, the interval that we would like to predict (short-term vs. long-term prognosis), the occurrence time of C, and finally the preceding history of the patient. This can, of course, be done only in the limits of the specified model and the data at hand.

In the BMT-example we would conclude, for example, that, given our model and data, the prediction of relapse within one year after transplantation of a male patient who has seropositive CMV-status, and who received a transplant after first remission from a female donor older than 20 years, and who had no mismatch between donor and patient cells, would change about 4% if he developed chronic GvHD within 6 months after transplantation, given that he had also had acute GvHD symptoms and CMV-infection previously. If he had not had acute GvHD, his prognosis of relapse would be about 13% better if he developed chronic GvHD (any time before one year after transplantation) than without it. On the other hand, if we were more interested in the short term prediction of relapse, we would conclude that such a patient's prognosis of relapse for six months after transplantation would improve only 1-2% due to chronic GvHD.

Furthermore, a patient with the same characteristics would "benefit" even 40% from chronic GvHD in the long run (in 360 days), if he had not had other complications (acute GvHD, CMV-infection) previously. Moreover, the probability that he would survive relapse-free at least a year after transplantation would increase about 43% if he developed chronic GvHD within 6 months after transplantation, given none of the other two complications previously, whereas his prognosis for relapse-free survival would not change, or would even slightly get worse, due to chronic GvHD if he already had both acute GvHD and CMV-infection. Such considerations are possible even though we could not specify the differential effects of the occurrence times of chronic GvHD and CMV-infection (if there are any).

A hypothetical example in which the occurrence times have a crucial role in determining the causal effect of an operation, was given in Arjas and Eerola (1993).

There several time-dependencies acted at the same time in the way that the conditional death hazards were assumed to depend on the time from prognosis (increasingly), on the age of the patient (increasing risk among older patients if treated), on the delay to operation (the longer the more), and on the time from operation (sudden increase in risk, but exponential decay). We could then conclude (using the hypothetical hazards) that the short-term effect (survival until age 40) of the operation was even harmful if the patient was older than age 25 when operated, whereas the long-term effect (survival until age 80) was clearly beneficial if the operation was performed before age 45.

In the 55-cohort data similar "individual" predictions can be given. In this example the dependencies between the events, although much discussed, are poorly understood, and the analysis is more or less exploratory. The effect of separation age is, however, a clearly time-related problem, and therefore a suitable test for the prediction method. Even though the number of subjects (n=3176) is large in these data, the event of interest is so rare that the number of subjects who experienced both separation and child psychiatric incidence was quite small (this is of course also an indication of fairly weak effects). In any case, the occurrence times of the events were in both cases stronger risk factors than the occurrence of the event as such. The difficulty, and also danger, when attempting to find parsimonious models, is that we inevitably oversimplify some dependencies. In the cohort data, for example, the specified model assumes that the earlier a child psychiatric incidence occurs, the more harmful it is. However, the observed range of T_C is [4,15] for those who had a subsequent hospitalization, and actually there are very few cases before age 12. The effect of T_C, condensed in the form $log(t - T_C)$, therefore extrapolates the values of the innovation gains outside of that range rather misleadingly. The effect of separation age was studied previously in Eerola (1989a,b) by time-dependent indicators. A slightly U-shaped effect was detected (in girls), such that age group $0 - 1$ had a larger coefficient than age group $1 - 3$. This could as well be a consequence of much smaller number of cases in the age group $1 - 3$ (cf. Fig. 4.16). Otherwise there was the same monotone increasing effect of separation age as was detected here.

Since regression coefficients usually, and also in this study, are very sensitive to a particular way of modelling, and particular models always are subjective statements about reality, we checked the sensitivity of the innovation gains in various ways. It is evident that likelihood-ratio tests are insufficient measures for this purpose. This has been pointed out also, for example, by Greenland and Robins (1986). In particular, they are useless when comparing the fit in dynamic models. We therefore used various checks for comparing the fit in time. The usual cautions, which apply for regression models, apply of course also here. Transformations which yield very deviating values at the extreme ranges of the variables should be avoided. A simple but useful way is to plot the curves using the estimated values of the coefficients. The overall fit of the model, both in the groups formed by the causal events and in total, can be checked by residual analysis where the residual groups are formed time-dependently, using the same principle as in time-dependent stratification. If we are analysing a particular causal chain, the main purpose may not, however, be the best possible fit of the model. If new

covariates are not included, the only possibility to improve the fit locally is to
redefine the parameterization of the underlying time (which includes all that has
not been modelled).

The innovation gains turned out to be much more stable measures of the causal
effects than the regression parameters. This conclusion bases, however, only on
one test data, and even there the other causal event (C) had a very weak effect
on the response. The differences due to varying $g(T_S)$ were larger, but even then
the innovation gains of different models were comparable unlike the regression
coefficients.

4.6. Computations

The estimation of parameters in the logistic regression models was done by a
nonstandard FORTRAN program (originally written by P. Kangas and R. Bloigu
from University of Oulu). An additional module was written and incorporated to
the program to calculate the martingale residuals in Chapter 4. The prediction
probabilities, innovation gains and the confidence limits in the examples of Chapter
4 were calculated by programs made for this purpose using the SAS-language,
so that all calculations were performed within a mainframe SAS system. The
probabilities of the hypothetical example in Chapter 2 were obtained by P.Kangas
by using the program Mathematica. All graphics was done by the SURVO84C-
system. Except for the confidence limits, the processing time needed to calculate
the prediction probabilities, and even the innovation gains, was reasonable. The
main determinants for the processing time are the number of events and the time
unit (the grid value) used. Nevertheless, a more efficient programming language
would be needed to calculate the confidence limits.

4.7 Further uses of the method

There are other potential uses of the prediction method which either general-
ize the event chain framework or otherwise fit to some well-known problems in
longitudinal studies. We will close this chapter with two such examples. In both
cases, we shall just outline some ideas and emphasize that further work is needed
in these areas.

4.7.1. Quantitative causal factors

If we simplify the question of causal effects as follows: "Does a change in X at
t cause a change in Y at $s > t$ in certain circumstances?", we have in this book
considered the case where X (and Y) are *events* and a change in X means the
occurrence of X at t. More generally, X, and of course also Y, can be *quantitative*,
and a change in X then means a change in the level of X at t. Such problems arise

frequently in medicine, for example, in the so called marker analysis where changes in the marker values affect the prediction of some disease (e.g. changes in the level of CD4 lymphocytes when predicting AIDS occurrence after HIV-infection (e.g. Galai, 1993)).

Obviously, the conditioning history then becomes overwhelmingly complicated if all possible paths of the processes X are considered. However, in many practical problems the number of actually interesting changes in the quantitative causal factors can be limited. Consider, for example, a risk factor Z, the normal values of which do not raise the risk of a disease (e.g. cholesterol level) but some harmful upper limit can be defined. One would then be interested in the events where such exceedings happen. There is usually some normal fluctuation in the values of the risk factors but the causally interesting interpretation is that only high/low values *cause* excess risk of the disease. In many cases the causal influence of such exceedings is not immediate but it is the cumulative staying in an excess risk state (e.g. high blood pressure) that eventually leads to the disease occurrence.

We shall now sketch two alternative ideas how quantitative risk factors could be handled in the framework of the prediction method. Consider a k-point causal chain in which the first $k-1$ events are assumed to cause the response C_k. Let (T, X) be a marked point process describing the evolution of this chain of events as before. Unlike before, we now assume that $C_1, ..., C_{k-1}$ are some causally meaningful events related to the quantitative risk processes $Z^1, ..., Z^{k-1}$. Given $\{T_n = s\}$, the mark $\{X_n = (i, r)\}, i = 1, ..., N, r = 1, ..., k$, identifies the individual and the component process for which the causally interesting event occurs at T_n.

(i) Fixed critical level. Assume first that for some Z^r there exists a fixed critical level z_r^* and the values which exceed this level are causally meaningful changes in Z^r. If (T, X) is considered as a multivariate counting process $\mathbf{N} = (N_s^r(i); r = 1, ..., k, i = 1, ..., N, t < s \leq u)$ registering such exceedings until the end of the prediction interval u, the rth component process for individual i is of the form

$$N_u^r(i) = \sum_{n \geq 1} 1_{\{T_n \leq u, X_n = (i,r)\}} = \sum_{s \leq u} 1_{\{T_{i,r} \leq s\}}$$

where

$$T_{i,r} = inf\{s : Z_s^r(i) > z_r^*\}, \text{ for } r = 1, ..., k-1,$$

and $T_{i,k}$ is the occurrence time of the response C_k.

The only difference, compared to the examples in Chapter 4, is that the dynamics of the rth risk factor $Z_s^r(i)$ in the prediction interval $(t, u]$ is now modelled by the hazard of the event $\{Z_s^r(i) > z_r^*\}, i = 1, ..., N, t < s \leq u$, which is straightforward. A natural question is, for example, the following: "What is the prediction of contracting a disease for an individual of type a whose value of the risk factor r exceed the level z_r^* at age t?". Conversely, how much would the prediction change if this value could be normalized by some age $t^* > t$?

(ii) Changing mean level. A more complicated situation arises if the aim is to predict the incidence of a disease in a population and the mean level of Z^r

varies in time. Then also the number of exceedings vary during the prediction interval, although it is of course possible that high risk individuals remain in the high risk group even if the mean level (e.g. cholesterol level) in the population decreases. If no critical level can be defined, a natural measure of excess risk is some deviation from the mean.

In this case, the event $\{T_n = s, X_n = (i,r)\}$ corresponds, for example, the event $\{Z^r_{T_n}(i) > \hat{\mu}^r_{T_n} + c(v\hat{a}r(\hat{\mu}^r_{T_n}))^{1/2}\}$, and an appropriate individual counting process is of the form

$$N^r_u(i) = \sum_{s \leq u} 1_{\{T_{i,r} \leq s\}}$$

where now

$$T_{i,r} = inf\{s : Z^r_s(i) > \hat{\mu}^r_s + c(v\hat{a}r(\hat{\mu}^r_s))^{1/2}\} \text{ for } r = 1, ..., k-1,$$

$T_{i,k}$ is the occurrence time of the response, and $\hat{\mu}^r_s$ the estimator of $\mu^r_s = E(Z^r_s|\mathcal{F}_t)$, $t \leq s$. The multiplier c is a constant denoting the extent of exceedings, and varies depending on what subgroups are considered.

To obtain the hazard of the event $\{Z^r_s(i) > \hat{\mu}^r_s + c(v\hat{a}r(\hat{\mu}^r_s))^{1/2}\}$, we need to estimate $\mu^r_s = E(Z^r_s|\mathcal{F}_t), t \leq s$ from the data. The conditional expectation of Z^r depends not only on its own history but on the history of all other risk factors as well. If the response event is recurrent, the number of previous relapses for an individual might change the conditional expectation of $Z^r_s(i)$ also. The conditioning history $(\mathcal{F}_s)$ is then typically of the form $\mathcal{F}_s = \mathcal{F}_0 \vee \mathcal{F}^Z_s \vee \mathcal{F}^{N^k}_s$ where $\mathcal{F}_0$ includes information of the initial conditions at t_0, $\mathcal{F}^Z_s = \sigma\{Z_t; t \leq s\}$; i.e., the history generated by the $(k-1)$-variate risk process, and $\mathcal{F}^{N^k}_s = \sigma\{N^k_t; t \leq s\}$ the history generated by the response process (the event C_k) until s.

If the dynamics of the risk factors can be simplified as in (i) or (ii), joint modelling of the response and the risk factors proceeds exactly as in the event chain models in Chapter 4. The conditional means in (ii) are estimated separately by specifying the dependence structure of the risk factors. Depending on their measurement level, different approaches are available. If the risk factors are measured continuously, or approximately so, the data could be regarded as a realization of a multivariate diffusion process (cf. Manton et al., 1989). The problem is then to estimate the drift and diffusion parameter of the component process Z^r at each $s \leq u$, given the history of all risk factors and $\sum_i 1_{\{T_{i,k} \geq s\}}$, the risk set at s if C_k is nonrecurrent. Manton et al. modelled the evolution of Z^r as a simple AR(1)-process and assumed that the joint transition function is multinormal at s. Obviously, the dependence on the history and on other risk factors can be much more complicated, and the order of the events may not be as easily interpreted as in event chains.

4.7.2. Informative censoring and drop-out

A common problem in follow-up studies is that some of the subjects in the study cannot be followed through the entire study period for various reasons. The usual assumption is then that such *censoring* or drop-out is uninformative for the estimation of the parameters of interest and its probability mechanism need not be modelled in the likelihood. The censoring times, even if actually random, are regarded as ancillary and, according to the conditionality principle, inference proceeds by conditioning on the observed censoring times.

In some occasions drop-out or censoring may, however, carry information about the causal dependence. Consider, for example, a job where the workers are exposed to a carcinogenic chemical which cumulatively adds the risk of cancer. Some workers leave the job (randomly in time) and the exposure then ceases. Sometimes it is possible to observe the response (in this case the incidence of cancer) also for the drop-outs. In order to measure the dose-response relationship between the chemical exposure and cancer incidence correctly, such short-term exposures should also be accounted for. In this example, censoring corresponds to leaving the job at time t and its informativeness is related to the length of exposure.

In the event chain framework censoring or drop-out can then be defined as one state of the chain from which there is a direct link to the response, cancer incidence. In terms of the prediction method, censoring at a particular time corresponds to the innovation gain of the exposure length when predicting cancer incidence. In experimental studies we can usually measure the dose until censoring exactly and entry times are the same for all individuals (e.g. time of randomization), but in general they can vary individually.

In such cases the prediction method offers a general framework in which the drop-out problem can be considered as a part of the whole prediction system. Its effect on the causal relationship can be evaluated dynamically at each timepoint in the prediction interval. This is a very natural approach especially in many dose-response studies in which the length of exposure corresponds often the amount of exposure.

Modelling the dynamics of censoring in the same way as in Chapter 4, i.e., by a parametric model, can be difficult because censoring is often caused by reasons which cannot be measured. Diggle and Kenward (1994) use, however, a logistic explanatory model for the drop-out process in a discrete repeated measurements situation where the drop-out probability depends on the history of the risk factors (e.g. the values of the processes Z_i^r in 4.7.1). Another approach to estimate the occurrence of censoring could be the use of some smoothing technique.

5. CONCLUDING REMARKS

The aim of this work was to present a method for the statistical analysis of causal effects in a series of events. One might think that this formulation of the task already presupposes that it is possible to confirm causal hypotheses by statistical methods (i.e., that causal statements can be verified by inductive inference). In philosophical discussion, and also in some applied fields, there is a firm belief among (Popperian) scientists that causal inference cannot be inductive in any phase, and a particular data set can only refute or corroborate a causal hypothesis. Therefore, observing n cases which support the hypothesis, no matter how large n is, will never confirm that a next observation would not contradict the hypothesis. Of course, in probabilistic causality we would not even require such constant conjunction of events. However, we do not even think that the role of statistical analysis in causality could be confirmative in any sense.

Causality is fundamental in all scientific inference and even in daily life our mind operates, more or less, in a causal way. As Hume already wrote, it is the empirical observation of conjoined occurrence of events that raises the idea of causal dependence between them, and a large part of scientific work relates to finding explanations for *why* they appear together. Statistical analysis can never answer to questions of this kind, but given hypothetical descriptions of how causal processes operate, statistical methods can help to make inferences about the amount of uncertainty (i.e., the sizes of causal effects) related to these hypothetical descriptions, whether this uncertainty is interpreted as ignorance of true causal mechanisms or inherent randomness in them. Statistical methods can also, as is the purpose of the prediction method suggested in this work, help to investigate whether such dependencies hold in variable circumstances, including time variation.

Although statistical modelling (experimentation) explicitly aims at separating effects of single factors and controlling for confounders, it seems that the notion of a causal field is particularly important in causal studies. We are also inclined to think that the concept of propensities serves for the same purpose in that it is the property of certain circumstances or "test" conditions to bring about certain outcomes rather than that of single events. In the causal chain framework these circumstances can be interpreted as the state of the chain at a particular moment. Consequently, the innovation gain from observing a particular event C_r at t can be interpreted as the causal effect of changing the prevailing circumstances by the occurrence of C_r. This also seems to be one of the differences between static and dynamic causal studies. In dynamic approaches the aim is to model a causal process, that is, the stages of a process in some simplified way. At a particular stage (i.e., in particular circumstances) a change can have different impact on the prediction of the response, and also the probability of the response itself, without any changes, varies in time. So, even though time is not a causal component itself, "true" causal components which evolve in time, can have different effects on the response in different circumstances.

In dynamic approaches we also consistently learn from the past in the sense that the innovation gain from observing an event at t is the unexpected new information which, when choosing $t = T_{C_r}$, corresponds to the causal effect of C_r at t on the event of prediction. If C_r had not occurred, the prediction of the response is determined by the past; i.e., the information of all that has happened before t and by the fact that nothing happened at t. In the prediction method we therefore explicitly state what is held fixed at t (the past) and what is let vary (the occurrence of the cause). In this sense, the method provides a "laboratory" framework to test various causal hypotheses within the considered causal chain. This construction of the innovation gain also closely resembles to measures suggested for the degree of confirmation of evidence (or information related to it) concerning a hypothesis.

Although strict objectivists consider propensities as determined by physical laws, and their existence independent of human consciousness, we would like to borrow from both objectivist and subjectivist sides. In most practical situations the only realistic interpretation is that all probabilities are dependent on the state of knowledge of those relevant "test" conditions. Even if we were able to determine the conditions in greatest detail, statistical models are always simplifications of reality to an extent which is decided subjectively by the researcher. However, this is not the fundamental dilemma for a statistician pondering on the problem of causal dependence in his or her models. More important questions are undoubtedly those raised by Rubin, Holland and others concerning the observational design. A question such as "What are the additional assumptions needed for causal inference when the observational plan does not correspond to the ideal design of a randomized experiment in which, at least in principle, the trials at present and in future are exchangeable?" is closely related to statistical practice. Although it is never possible to be certain that there is no confounding in the study design (and this is again an epistemic question), it is important to realize the restrictions on causal inference in statistical analysis. It is clear that the prediction method is as sensitive as any other statistical method to bias due to confounding or non-random samples.

The possibility to make inference on 'true' unit level causal relationships is of course dependent on our ability to control for those relevant factors which could distort the unit-treatment effect; i.e., given those factors, the influence of a cause would be the same on each unit. Obviously, there is an endless amount of differentiating features among, for example, human beings, and therefore finding the relevant ones in a particular situation can never be decided by statistical methods.

Another, more technical question related to statistical modelling concerns the particular form of analysis: nonparametric or parametric. Roughly speaking, in nonparametric analysis a particular data set shows "how reality works". In parametric analysis, by postulating a statistical model, for example, for the hazard, we make hypotheses about reality in a more general sense. This can, of course, be completely misleading if these hypotheses do not fit the data. Usually the hypotheses are, or at least they should be, implications of some a priori knowledge or understanding about the problem at hand. On the other hand, if the hypotheses are realistic, we can often gain more insight into the problem.

If a completely general dependence structure in a causal chain (or, in general in a series of random events) is allowed for, then nonparametric analysis is practically impossible. Dependence on the history is expressed in the model by stochastic covariates, and their evolution in time must be determined also. As in the examples of Chapter 4, various hypotheses about the dependence structure can be tested. It was demonstrated in various ways that the analysis can be misleading if the realism of these hypotheses is not checked. This is, of course, not just the problem of the prediction method but concerns all statistical modelling. Several assumptions are usually implicitly made in statistical models. When such assumptions, both statistical and causal, are made, the consequences of these assumptions should always be checked by careful sensitivity analysis.

Finally, the usefulness of the proposed prediction method relies obviously very much on the researcher's ability to express his or her knowledge about the causal processes in a simplified but meaningful way as a chain of events. The variety of other factors altering the causal effects can be taken into account in the hazard models as in usual regression analysis. Obviously, such detailed expressions of dependencies between the occurrence times are not always necessary. Sometimes it is sufficient to know that the hypothetical cause occurred by time t, no matter when exactly. Even then the setup of the proposed method is useful for causal (and non-causal) analysis of dependent events in that it explicitly measures the effect of change in the prevailing conditions, caused by the considered event, on the probability of the response. In our view, this is the natural interpretation of probabilistic statements about causal effects of events.

BIBLIOGRAPHY

Aalen, O. (1976) Statistical theory for a family of counting processes. Institute of Mathematical Statistics, University of Copenhagen.

Aalen, O. (1978) Nonparametric estimation of partial transition probabilities in multiple decrement models. Ann. Statist. 6: 534-45.

Aalen, O. (1987) Dynamic modelling and causality. Scand. Actuarial J. 177-190.

Aalen, O., Borgan, O., Keiding, N. & Thorman, J. (1980) Interaction between life-history events. Nonparametric analysis of prospective and retrospective data in the presence of censoring. Scand. J. Statist. 7: 161-171.

Aalen, O. & Johansen, S. (1978) An empirical transition matrix for non-homogenous Markov chains based on censored observations. Scand.J.Statist. 5: 141-50.

Almqvist, F. (1983) Psykiatriska vårdkontakter och registrerad social missanpassning under ålderperioden 15-21 år. Kansanterveystieteen laitoksen julkaisuja M 72. Helsinki.

Amnell, G. (1974) Mortalitet och kronisk morbiditet i barnåldern. Samfundet Folkhälsan, Helsinki.

Andersen, P.K. (1986) Time-dependent covariates and Markov processes. In: Moolgavkar, S. H. & Prentice, R.L. (eds) Modern Statistical Methods in Chronic Diseases Epidemiology. Wiley, New York.

Andersen, P.K. (1988) Multi-state models in survival analysis: a study on nephropathy and mortality in diabetes. Statist. in Medicine 7: 661-670.

Andersen, P.K. & Borgan, O. (1985) Counting process models for life-history data: a review. Scand. J. Statist. 12: 97-158.

Andersen, P.K., Borgan, O., Gill, R.D. & Keiding, N. (1988) Censoring, truncation and filtering in statistical models based on counting processes. Contemp. Math. 80: 19-60.

Andersen, P.K. & Gill, R.D. (1982) Cox's regression model for counting processes: a large sample study. Ann. Statist. 10: 1100-1120.

Andersen, P.K., Hansen, L.S. & Keiding, N. (1991) Non- and semiparamtèric estimation of transition probabilities from censored observation of a non-homogenous Markov process. Scand. J. Statist. 18: 153-164.

Andersen, P.K. & Rasmussen, N.K. (1982) Admissions to psychiatric hospitals among women giving birth and women having induced abortion. A statistical analysis of a counting process model. Statist. in Medicine 5: 243-253.

Arjas, E. (1986) Standford heart transplantation data revisited: A real time approach. In: Moolgavkar, S. H. & Prentice, R.L. (eds) Modern Statistical Methods in Chronic Diseases Epidemiology. Wiley, New York.

Arjas, E. (1988) A graphical method for assessing goodness-of-fit in Cox's regression model. J. Am. Statist. Assoc. 83: 204-212.

Arjas, E. (1990) Survival models and martingale dynamics. Scand. J. Statist. 16: 177-225

Arjas, E. & Eerola, M. (1993) On predictive causality in longitudinal studies. J. of Statist. Planning and Inference 34: 361-386.

Arjas, E. & Haara, P. (1984) A marked point process approach to censored failure time data with complicated covariates. Scand. J. Statist. 11: 193-209.

Arjas, E. & Haara, P. (1985) A note on the exponentiality of the total hazards before failure. J. Multivariate Anal. 26: 207-218.

Arjas, E. & Haara, P. (1986) A logistic regression model for hazards. Scand. J. Statist. 14: 1-18.

Arjas, E. & Norros, I. (1984) Life lengths and association: a dynamic approach. Math. Oper. Res. 9: 151-158.

Arjas, E. & Norros, I. (1991) Stochastic order and martingale dynamics in multivariate life length models: a review. In: Mosler, R. & Scarsini, M. (eds) Stochastic order and decision under risk. IMS Lecture Notes, Monograph Series 19, 7-24.

Armitage, P. & Doll, R. (1954) A two-stage theory of carcinogenesis in relation to the age distribution of human cancer. Br. J. Cancer 11: 161-169.

Barlow, W.E. & Prentice, R.L. (1988) Residuals for relative risk regression. Biometrika 75: 65-74.

Blalock, H.M. (1971) Causal Models in the Social Sciences. Aldine-Atherton, Chigago.

Bowlby, J. (1969) Attachment and Loss. Vol 2: Separation. Hogart Press, London.

Brand, M. (1976) The Nature of Causation. The Univ. of Illinois Press, Illinois.

Brèmaud, P. (1981) Point processes and queues. Martingale dynamics. Springer-Verlag, New York.

Breslow, N. & Crowley, J. (1974) A large sample study of the life table and product limit estimates under random censorship. Ann. Statist. 2: 437-453.

Cain, K.C. & Lange, N.T. (1984) Approximate case influence for the proportional hazards regression model with censored data. Biometrics 40: 493-499.

Casella, G. & Strawderman, W. (1980) Confidence bands for linear regression with restricted predictor variables. J. Am. Statist. Assoc. 75: 862-868.

Chamberlain, G. (1982) The general equivalence of Granger and Sims causality. Econometrica 50: 569-580.

Chiang, C.L. (1968) Introduction to Stochastic Processes in Biostatistics. Wiley, New York.

Cox, D.R. (1992) Causality: some statistical aspects. J. R. Stat. Soc. A, 155: 291-302.

Cox, D. & Wermuth, N. (1993) Linear dependencies represented by chain graphs (with discussion). Statistical Science 8, 204-218.

Crowley, J. & Storer, B.E. (1983) Contribution to the discussion on the paper by Aitkin, Laird and Francis. J. Am. Statist. Assoc. 78: 277-281.

Davis, W. (1988) Probabilistic theories of causation. In: Fetzer, J. (ed) Probability and Causality. D. Reidel, Dordrecht, pp. 133-160.

Dempster, A. (1990) Causality and statistics. J. Statist. Planning and Inference 25: 261-278.

Diggle, P. & Kenward, M. (1994) Informative drop-out in longitudinal data. Applied Statistics 43, 49-94.

Ducasse, H. (1951) Critique of Hume's analysis. In: La Salle (ed) Nature, Mind and Death. Open Court Publ., Illinois, pp. 91-100.

Eells, E. & Sober, E. (1983) Probabilistic causality and the question of transitivity. Philos. of Science 50: 33-57.

Eerola, M. (1989a) Repeatable events in event-history analysis; the effect of childhood separation on subsequent mental hospitalisations. Unpublished Licentiate Thesis. University of Helsinki.

Eerola, M. (1989b) Repeatable events in event-history analysis; the effect of childhood separation on future mental health. Bull. Int. Statist. Inst. vol. LIII(1), 213-225.

Efron, B. (1988) Logistic regression, survival analysis, and the Kaplan-Meier curve. J. Am. Statist. Assoc. 83: 414-425.

Eskola, A. (1989) Tulevaisuuden ennusteet ja lainmukaisuus sosiologiassa. In: Heiskanen, P. (ed.) Aika ja sen ankaruus. Gaudeamus, Helsinki, pp. 185-191.

Fetzer, J. (1974) Statistical probabilities: Single-case propensities vs. long-run frequencies. In: Leinfellner, W. & Kohler, E. (eds) Developments in the Methodology of Social Science. D. Reidel, Dordrecht,pp. 387-397.

Fetzer, J. (1988) Probabilistic metaphysics. In: Fetzer, J. (ed.) Probability and Causality. D. Reidel, Dordrecht,pp. 133-160.

de Finetti, B. (1972) The Theory of Probability. Wiley, New York.

Fisher, R. A. (1934) The Design of Experiments. Oliver and Boyd, Edinburgh.

Fleming, T. & Harrington, D. (1991) Counting Processes and Survival analysis. Wiley, New York.

Florens, J. & Mouchart, M. (1982) A note on non-causality. Econometrica 50: 583-591.

Gail, M.H., Santner, T.J. & Brown, C.C. (1980) An analysis of comparative carcinogenesis experiments based on multiple times to tumor. Biometrics 36: 255-266.

Galai, N., Munoz, A., Chen, K., Carey, V., Chmiel, J. & Zhou S. (1993) Tracking of markers and onset of disease among HIV-1 seroconverters. Statistics in Medicine 12, 2133-2145.

Giere, R. (1973) Objective single-case probabilities and the foundations of statistics. In: Suppes, P., Henkin, L. Moisil, Gr. & Joja, A. (eds). Logic, Methodology, and Philosophy of Science, IV. North-Holland, Amsterdam, pp.467-483.

Gill, R.D. (1980) Censoring and Stochastic Integrals. MC Tract 124, Mathematisch Zentrum, Amsterdam.

Gill, R.D. & Johansen, S. (1987) A survey of product integrals with a view towards application in survival analysis. Ann. Statist. 18: 1501-55.

Good, I.J. (1961/62) A causal calculus I-II. Brit. J. Philos. Science 11: 305-318 and 12: 43-51.

Good, I.J. (1988) Causal tendency: a review. In: Skyrms, B. & Harper, W. (eds) Causation, Chance and Credence. Kluwer Academic Publ., Dordrecht, pp. 23-50.

Granger, C.W.J. (1969) Investigating causal relations by econometric models and cross-spectral methods. Econometrica 37: 424-438.

Granger, C.W.J. (1986) Comment on the paper by P. Holland. J. Am. Statist. Assoc. 81: 967-968.

Greenland, S. (1990) Randomization, statistics, and causal inference. Epidemiology 1: 421-428.

Greenland, S. & Robins, J. (1990) Identifiability, exchangeability, and epidemiologic confounding. Int. J. Epidemiol. 15: 413-419.

Hacking, I. (1965) The Logic of Statistical Inference. Cambridge University Press, Cambridge.

Hall, P. & Wellner, J. (1980) Confidence bands for a survival curve from censored data. Biometrika 67: 133-143.

Hoem, J. (1985) Weighting, missclassification and other issues in the analysis of survey samples of life histories. In: Heckman, J. & Singer, B. (eds) Longitudinal Analysis of Labor Market Data, Chapter 7. Cambridge University Press, Cambridge.

Holland, P. (1986) Statistics and causal inference (with discussion) J. Am. Statist. Assoc. 81: 945-970.

Holland, P. (1989) Comment on the paper of Wickramaratne, P. and Holford, T. Biometrics 45: 1310-1316.

Hume, D. (1741) An Inquire Concerning Human Understanding (ed. C. Hendell). The Liberal Arts Press, New York.

Hume, D. (1748) Treatise of Human Nature. Oxford Univ. Press, Oxford. (1978, text revised by P. Nidditsch).

Humphreys, P. (1985) Why propensities cannot be probabilities? Philosophical Review 94: 557-570.

Jacobsen, M. (1982) Statistical Analysis of Counting Processes. Lecture Notes in Statistics 12. Springer-Verlag, Berlin.

Jacobsen, N., Badsberg, J., Lönnqvist, B., Ringden, O., Volin, L., Rajantie, J., Koskelainen, J. & Keiding, N. (1990) Graft-versus-leukemia activity associated with CMV-seropositive donor, post-transplant CMV infection, young donor age and chronic graft-versus-host disease in bone marrow allograft recipients. Bone marrow transplantation 5: 413-418.

Jöreskog, K. & Sörbom, D. (1977) Statistical models and methods for analysis of longitudinal data. In: D. Aigner & A. Goldberger (eds) Latent Variables in Socio-economic Models, pp. 285-325.

Kalbfleisch, J.D. & Lawless, J. (1988) Likelihood analysis of multi-state models for disease incidence and mortality. Statist. in Medicine 7: 149-160.

Kant, I. (1974) Logic. Boobs-Merril, New York (orig. 1800).

Kant, I. (1789) Kritik der Reinen Vernunft.

Keiding, N. & Andersen, P.K. (1989) Nonparametric estimation of transition intensities and transition probabilities: a case study of a two-state Markov process. Appl. Statist. 38: 319-329.

Kiiveri, H. & Speed, T. (1982) Structural analysis in multivariate data: a review. In: S. Leinhardt (ed). Sociological Methodology. Jossey-Bass, San Francisco, 209-289.

Klein, J.P., Keiding, N. & Copelan, E.A. (1994) Plotting summary predictions in multistate survival models: Probabilities of relapse and death in remission for bone marrow transplantation patients. Statist. in Medicine 12: 2315-2332.

Kyburg, H. (1970) Probability and Inductive Logic. MacMillan, London.

Laplace, P.S. (1774) Memoire sur la probabilité des causes par les évènement. Translated by Stigler, S. in Statistical Science (1986), vol 1.

Lauritzen, S. & Wermuth, N. (1989) Graphical models for associations between variables some of which are qualitative and some quantitative. Ann. Statist. 17: 31-57.

Lewis, D. (1973) Counterfactuals. Blackwell, Oxford.

Longini, I., Clark, W. & Haber, M. (1989) The stages of HIV-infection: waiting times and infection transmission probabilities. In: C. Cartillo-Chevez (ed.) Mathematical and Statistical Approaches to AIDS-epidemic. Lecture Notes in Biomathematics 83: 111-137. Springer-Verlag, Berlin.

Lönnqvist B., Ringden, O. & Ljungman, P. (1986) Reduced risk of recurrent leukemia in bone marrow transplant recipients after cytomegalovirus infection. Br J Haematol. 63: 671-679.

Mackie, J. (1973) The Cement of the Universe: A Study of Causation. Oxford University Press, Oxford.

Manton, K., Woodbury, M. & Stallard, E. (1988) Models of the interaction of mortality and the evolution of risk factor distribution: A general stochastic process formulation. Statist. in Medicine 7: 239-256.

Matthews, D. (1988) Likelihood-based confidence intervals for functions of many parameters. Biometrika 75: 139-144.

Mellor, D. (1988) On raising the chances of effects. In: Fetzer, J. (ed.) Probability and Causality. D. Reidel, Dordrecht, pp. 133-160. 109-131.

Miettinen, O. (1982) Causal and preventive interdependence: elementary principles. Scand. J. Work, Environment and Health 8: 159-168.

Mill, J.S. (1843) A system of Logic. J.V. Parker, London.

von Mises, R. (1928) Probability, Statistics and Truth. Allen and Unwin, London.

Moolgavkar, S. & Knudson, A. (1981) Model for human carcinogenesis. J. Nat. Cancer Inst. 6: 1037-1051.

Mykland, P. (1986) Statistical causality. Technical report. Dept. of Math. Univ. of Bergen, Bergen.

Nagel, E. (1961) The Structure of Science: Problems in the Logic of Scientific Explanation. Routledge & Kegan Paul, London.

Nair, V. (1984) Confidence bands for survival functions with censored data; A comparative study. Technometrics 14: 945-965.

Norros, I. (1985) System weakened by failures. Stochast. Process. Appl. 20: 181-196.

Norros, I. (1986) A compensator representation of multivariate life length distributions, with applications. Scand. J. Statist. 13: 99-112.

Pepe, M.S. (1991) Inference for events with dependent risks in multiple endpoint studies. J. Am. Statist. Assoc. 86: 770-778.

Pepe, M.S., Longton, G. & Thornquist, M. (1991) A qualifier Q for the survival function to describe the prevalence of a transient condition. Statist. in Medicine 10: 413-421.

Piegorsch, W. & Casella, G. (1988) Confidence bands for logistic regression with restricted predictor variables. Biometrics 44: 739-750.

Prentice, R., Kalbfleisch, J., Peterson, A., Flournoy, N., Farewell, V. & Breslow, N. (1978) The analysis of failure time data in the presence of competing risks. Biometrics 34: 541-554.

Popper, K. (1959) The propensity interpretation of probability. British Journal of Philosophy of Science 10: 26-42.

Ramsey, F. (1926) Truth and Probability. In: Braithwaite, R. (ed.) The Foundations of Mathematics. Routledge & Kegan Paul, London.

Reichenbach, H. (1949) The Theory of Probability. Univ. of California Press, Berkeley.

Robins, J. (1989) The control of confounding by intermediate variables. Statist. in Medicine 6: 679-702.

Robins, J. (1992) Estimation of the time-dependent accelerated failure time model in the presence of confounding factors. Biometrika 79: 321-334.

Rosenbaum, P. & Rubin, D. (1983) The central role of propensity score in observational studies. Biometrika 70: 41-55.

Rothman, K., Greenland, S. & Walker, A. (1980) Concept of interaction. Am. J. Epidemiol. 112: 467-470.

Rubin, D. (1974) Estimating causal effects of treatments in randomized and nonrandomized studies. J. Educ. Psychol. 66: 688-701.

Rubin, D. (1978) Bayesian inference for causal effects; the role of randomization Ann. Statist. 6: 34-58.

Rubin, D. (1990) Formal modes of statistical inference for causal effects. J. Statist. Planning and Inference 25: 279-292.

Salmon, W. (1980) Probabilistic causality. Pacific Philos. Quart. 61: 50-74.

Salmon, W. (1984) Scientific Explanation and the Causal Structure of the World. Princeton Univ. Press, Princeton.

Salmon, W. (1988) Dynamic rationality: propensity, probability and credence. In: Fetzer, J. (ed.) Probability and Causality. D. Reidel, Dordrecht, pp. 109-131.

Savage, L. (1954) The Foundations of Statistics. Dover Publ., London.

Scheffé, H.(1953) A method for judging all contrasts in the analysis of variance. Biometrika 40: 87-104.

Schoenfeld, D. (1982) Partial residuals for the proportional hazards regression model. Biometrika 69: 239-241.

Schweder, T. (1970) Composable Markov Processes. J. Appl. Prob. 7: 400-410.

Shaked, M. & Shanhtikumar, G. (1987) The multivariate hazard construction. Stochast. Proc. Appl. 24: 241-258.

Skyrms, B. (1988) Conditional Chance. In: Fetzer, J. (ed.) Probability and Causality. D. Reidel, Dordrecht, pp. 161-178.

Skyrms, B. (1988) Probability and causation. In: Skyrms, B. & Harper, W. (eds) Causation, Chance and Credence. Kluwer Academic Publ., Dordrecht, pp. 109-131.

Sosa, E. (1975) Causation and Conditionals. Oxford Univ. Press, Oxford.

Spirtes, P., Glymour, G. & Scheines, R. (1993) Causation, Prediction, and Search. Lecture Notes in Statistics 81. Springer-Verlag, Berlin.

Stalnaker, R. (1968) A theory of conditionals. In: Rescher, N. (ed.) Studies in Logical Theory. Blackwell, Oxford, pp. 98-112.

Suppes, P. (1970) A Probabilistic Theory of Causality. (Acta Philosophica Fennica 24). North-Holland, Amsterdam.

Suppes, P. (1984) Probabilistic Metaphysics. Blackwell, Oxford.

Suppes, P. (1986) Non-markovian causality in the social sciences with some theorems on transitivity. Synthese 68: 129-140.

Suppes, P. (1987) Some further remarks on propensity: reply to Maria Carla Galavotti. Erkenntnis 26: 369-370.

Suppes, P. (1990) Probabilistic causality in space and time. In: Skyrms, B. & Harper, W. (eds) Causation, Chance and Credence. Kluwer Academic Publ., Dordrecht, pp. 135-151.

Tennant, C., Smith, A., Bebbington, B. & Hurry, J. (1981) Parental loss in childhood: relationships to adult psychiatric impairment and contact with psychiatric services. Arch. Gen. Psychiatry 38: 309-314.

Therneau, T., Grambsch, P., Fleming, T. (1990) Martingale-based residuals for survival models. Biometrika 77: 147-160.

Tuma, N.B. & Hannan, M.T. (1984) Social Dynamics: Models and Methods. Academic Press, New York.

Weiden, P., Sullivan, K.,Flournoy, N., Storb, R., Thomas, E. (1981) Antileukemic effect of chronic-graft-versus-disease. New England J. Medicine 304: 1529-1533.

Venn, J. (1886) The Logic of Chance. MacMillan, London.

Wermuth, N. & Lauritzen, S. (1990) On substantive research hypotheses, conditional independence graphs and conditional chain models (with discussion). J. R. Stat. Soc. B 52: 21-72.

Whittaker, J. (1990) Graphical Models in Applied Multivariate Analysis. Wiley, Chichester.

Wiener, N. (1956) The theory of prediction. In: Beckenbach, E. (ed.) Modern Mathematics for the Engineer. McGraw-Hill, New York, pp. 165-190.

Working, H. & Hotelling, H. (1929) Application of the theory of error to the interpretation of trends. J. Am. Stat. Assoc. Suppl. 24: 73-85.

Wright, J. (1934) The method of path coefficients. J. Am. Math. Stat. 5: 161-215.

Väisänen, E. (1975) Mielenterveyden häiriöt Suomessa. Kansaneläkelaitoksen julkaisuja AL: 2. Helsinki.

Yarrow, L.J. (1964) Separation from parents during early childhood. In: Review of Child Development Research. Vol 1. Hoffman & Hoffman, New York.

Yule, G. (1903) Notes on the theory of association of attributes in statistics. Biometrika 2: 121-134.

Zwaan, F.E., Herman, J., Barret, A.J. & Speck B. (1984) Bone marrow transplantation for acute nonlymphoplastic leukaemia: a survey of the European group for bone marrow transplantation. Br. J. Haematol. 56: 645-653.

APPENDIX 1. DERIVATIVES OF THE PREDICTION PROBABILITIES

Hazard for event $x = (A, B, C)$ with the corresponding parameters $\theta = (\alpha, \beta, \gamma)$ in the logistic model:

$$p_x(t) = P(\Delta N_t(x) = 1|\mathcal{F}_{t-1}) = Y_x(t-1)\frac{exp(\sum_i \theta_i Z_i(t-1))}{1 + exp(\sum_i \theta_i Z_i(t-1))}.$$

Partial derivative of $p_x(t)$ w.r.t. to θ_j:

$$\frac{\partial p_x(t)}{\partial \theta_j} = Z_j(t-1)exp(-\sum_i \theta_i Z_i(t-1))p_x^2(t) = Z_j(t-1)p_x(t)(1 - p_x(t))$$

Partial derivative of $P(T_C > s|\mathcal{F}_t) = F_s(\theta); t < s \leq u$ w.r.t. θ_j is obtained by the chain rule:

$$F'_u(\theta) = \sum_{s=t+1}^{u} \frac{\partial F_s(\theta)}{\partial \theta}$$

$$= \sum_{s=t+1}^{u} \sum_{x} \frac{\partial F_s(\hat{\theta})}{\partial p_x(s)} \cdot \frac{\partial p_x(s)}{\partial \theta_j}; \quad \theta = (\alpha_1, ..., \alpha_{p_1}, \beta_1, ..., \beta_{p_2}, \gamma_1, ..., \gamma_{p_3}).$$

so that the estimated asymptotic covariance of $F_r(\hat{\theta})$ and $F_s(\theta), t < r \leq s \leq u$, is

$$\widehat{cov}(F_r(\hat{\theta}), F_s(\hat{\theta})) = \sum_{k=t+1}^{r} \sum_{l=t+1}^{s} F'_k(\hat{\theta}) I(\hat{\theta})^{-1} (F'_l(\hat{\theta}))'$$

Partial derivatives of $P(T_C > s|\mathcal{F}_t) = F_s(\theta), t < s \leq u$ w.r.t to θ_j with different history sets: For $v < w \leq t < s \leq u$

1) History set $\{(A, v), (B, w)\}$ at t:

$$\frac{\partial P(T_C > u|T_A = v, T_B = w, T_C > t)}{\partial \gamma_j} = -\sum_{s=t+1}^{u} Z_j(s-1)p_{C|AB}(s|v,w)(1 - p_{C|AB}(s|v,w)) \prod_{\substack{r=t+1 \\ r\neq s}}^{u} (1 - p_{C|AB}(r|v,w))$$

$$= -S_{C|AB}(u|v,w) \sum_{s=t+1}^{u} Z_j(s-1)p_{C|AB}(s|v,w)$$

(A.1)

2) History set $\{(B, w)\}$ at t:

$$\frac{\partial P(T_C > u|T_A = \infty, T_B = w, T_C > t)}{\partial \gamma_j} = -\sum_{s=t+1}^{u} Z_j(s-1)p_{C|B}(s|w)(1 - p_{C|B}(s|w)) \prod_{\substack{r=t+1 \\ r\neq s}}^{u} (1 - p_{C|B}(r|w)) \quad (A.2)$$

3) History set $\{(A, v)\}$ at t:

$$\frac{\partial P(T_C > u|T_A = v, T_B > t, T_C > t)}{\partial \beta_j} = -\sum_{s=t+1}^{u} Z_j(s-1)p_{B|A}(s|v)(1 - p_{B|A}(s|v)) \prod_{\substack{r=t+1 \\ r\neq s}}^{u} (1 - p_{B|A}(r|v) - p_{C|A}(r|v))$$

$$+ \sum_{s=t+1}^{u} Z_j(s-1) p_{B|A}(s|v)(1-p_{B|A}(s|v)) \prod_{r=t+1}^{s-1} (1-p_{B|A}(r|v) - p_{C|A}(r|v))$$
$$\times P(T_C > u | T_A = v, T_B = s, T_C > s)$$

$$- \sum_{s=t+1}^{u} [\sum_{r=t+1}^{s-1} Z_j(r-1) p_{B|A}(r|v)(1-p_{B|A}(r|v)) \prod_{\substack{q=t+1\\q\neq r}}^{s-1} (1-p_{B|A}(q|v) - p_{C|A}(q|v))] p_{B|A}(s|v) \quad (A.3)$$
$$\times P(T_C > u | T_A = v, T_B = s, T_C > s)$$

$$\frac{\partial P(T_C > u | T_A = v, T_B > t, T_C > t)}{\partial \gamma_j} = - \sum_{s=t+1}^{u} Z_j(s-1) p_{C|A}(s|v)(1-p_{C|A}(s|v)) \prod_{\substack{r=t+1\\r\neq s}}^{u} (1-p_{B|A}(r|v) - p_{C|A}(r|v))$$

$$- \sum_{s=t+1}^{u} [\sum_{r=t+1}^{s-1} Z_j(r-1) p_{C|A}(r|v)(1-p_{C|A}(r|v)) \prod_{\substack{q=t+1\\q\neq r}}^{s-1} (1-p_{B|A}(q|v) - p_{C|A}(q|v))] p_{B|A}(s|v)$$
$$\times P(T_C > u | T_A = v, T_B = s, T_C > s)$$

$$- \sum_{s=t+1}^{u} p_{B|A}(s|v) \prod_{r=t+1}^{s-1} (1-p_{B|A}(r|v) - p_{C|A}(r|v)) \frac{\partial P(T_C > u | T_A = v, T_B = s, T_C > s)}{\partial \gamma_j} \quad (A.4)$$

4) History set $\{\emptyset\}$ at t :

$$\frac{\partial P(T_C > u | T_A > t, T_B > t, T_C > t)}{\partial \alpha_j} = - \sum_{s=t+1}^{u} Z_j(s-1) p_A(s)(1-p_A(s)) \prod_{\substack{r=t+1\\r\neq s}}^{u} (1-p_A(r) - p_B(r) - p_C(r))$$

$$+ \sum_{s=t+1}^{u} Z_j(s-1) p_A(s)(1-p_A(s)) \prod_{r=t+1}^{s-1} (1-p_A(r) - p_B(r) - p_C(r))$$
$$\times P(T_C > u | T_A = s, T_B > s, T_C > s)$$

$$- \sum_{s=t+1}^{u} [\sum_{r=t+1}^{s-1} Z_j(r-1) p_A(r)(1-p_A(r)) \prod_{\substack{q=t+1\\q\neq r}}^{s-1} (1-p_A(q) - p_B(q) - p_C(q))] p_A(s)$$
$$\times P(T_C > u | T_A = s, T_B > s, T_C > s)$$

$$- \sum_{s=t+1}^{u} [\sum_{r=t+1}^{s-1} Z_j(r-1) p_A(r)(1-p_A(r)) \prod_{\substack{q=t+1\\q\neq r}}^{s-1} (1-p_A(q) - p_B(q) - p_C(q))] p_B(s) \quad (A.5)$$
$$\times P(T_C > u | T_A = \infty, T_B = s, T_C > s)$$

$$\frac{\partial P(T_C > u | T_A > t, T_B > t, T_C > t)}{\partial \beta_j} = - \sum_{s=t+1}^{u} Z_j(s-1) p_B(s)(1-p_B(s)) \prod_{\substack{q=t+1 \\ r \neq s}}^{u} (1 - p_A(r) - p_B(r) - p_C(r))$$

$$- \sum_{s=t+1}^{u} [\sum_{r=t+1}^{s-1} Z_j(r-1) p_B(r)(1-p_B(r)) \prod_{\substack{q=t+1 \\ q \neq r}}^{s-1} (1 - p_A(q) - p_B(q) - p_C(q))] p_A(s)$$
$$\times P(T_C > u | T_A = s, T_B > s, T_C > s)$$

$$- \sum_{s=t+1}^{u} p_A(s) \prod_{r=t+1}^{s-1} (1 - p_A(r) - p_B(r) - p_C(r)) \frac{\partial P(T_C > u | T_A = s, T_B > s, T_C > s)}{\partial \beta_j}$$

$$+ \sum_{s=t+1}^{u} Z_j(s-1) p_B(s)(1-p_B(s)) \prod_{r=t+1}^{s-1} (1 - p_A(r) - p_B(r) - p_C(r))$$
$$\times P(T_C > u | T_A = \infty, T_B = s, T_C > s)$$

$$- \sum_{s=t+1}^{u} [\sum_{r=t+1}^{s-1} Z_j(r-1) p_B(r)(1-p_B(r)) \prod_{\substack{q=t+1 \\ q \neq r}}^{s-1} (1 - p_A(r) - p_B(r) - p_C(r))] p_B(s) \quad (A.6)$$
$$\times P(T_C > u | T_A = \infty, T_B = s, T_C > s)$$

$$\frac{\partial P(T_C > u | T_A > t, T_B > t, T_C > t)}{\partial \gamma_j} = - \sum_{s=t+1}^{u} Z_j(s-1) p_C(s)(1-p_C(s)) \prod_{\substack{r=t+1 \\ r \neq s}}^{u} (1 - p_A(r) - p_B(r) - p_C(r))$$

$$- \sum_{s=t+1}^{u} [\sum_{r=t+1}^{s-1} Z_j(r-1) p_C(r)(1-p_C(r)) \prod_{\substack{q=t+1 \\ q \neq r}}^{s-1} (1 - p_A(q) - p_B(q) - p_C(q))] p_A(s)$$
$$\times P(T_C > u | T_A = s, T_B > s, T_C > s)$$

$$- \sum_{s=t+1}^{u} p_A(s) \prod_{r=t+1}^{s-1} (1 - p_A(r) - p_B(r) - p_C(r)) \frac{\partial P(T_C > u | T_A = s, T_B > s, T_C > s)}{\partial \gamma_j}$$

$$- \sum_{s=t+1}^{u} [\sum_{r=t+1}^{s-1} Z_j(r-1) p_C(r)(1-p_C(r)) \prod_{\substack{q=t+1 \\ q \neq r}}^{s-1} (1 - p_A(q) - p_B(q) - p_C(q))] p_B(s)$$
$$\times P(T_C > u | T_A = \infty, T_B = s, T_C > s)$$

$$- \sum_{s=t+1}^{u} p_B(s) \prod_{r=t+1}^{s-1} (1 - p_A(r) - p_B(r) - p_C(r)) \frac{\partial P(T_C > u | T_A = \infty, T_B = s, T_C > s)}{\partial \gamma_j} \quad (A.7)$$

APPENDIX 2: RESULTS OF ESTIMATED HAZARD MODELS

HAZARD OF CMV-INFECTION   N=163   (BONE MARROW TRANSPLANTATION DATA)

****
* ESTIMATED MODEL *   GRID INTERVAL :   1.000000
*********************   -2 * LOG (LIKELIHOOD) :   1258.975

PARAMETER NUMBER	COVAR. INDEX	COVARIATE NAME	PARAMETER ESTIMATE	EST STD DEV	EXP(PARAMETER)	PARAM/ST.ERR.
FIXED COVARIATES :						
1	0	CONSTANT	-6.0662	0.2967	2.320E-03	-20.4460
2	14	patient cmv-status	1.4258	0.2572	4.1611	5.5433
3	15	mismatch	0.4309	0.2867	1.5386	1.5026
4	18	donor age (>20)	0.1431	0.2132	0.8666	-0.6713
PRESET TIME DEPENDENT COVARIATES :						
5	1	1(T_A<t)	0.8299	0.2074	2.2932	4.0000
COMPUTED TIME DEPENDENT COVARIATES :						
6	1	Z24 (t) = 1(t>90)	-1.3950	0.2538	0.2478	-5.4951

*** ESTIMATED ASYMPTOTIC CORRELATION MATRIX OF PARAMETERS ***

	1	2	3	4	5	6
1	1.0000					
2	-0.8150	1.0000				
3	-0.2835	0.1621	1.0000			
4	-0.5210	0.2780	0.2199	1.0000		
5	-0.2571	-0.0521	-0.1608	0.0342	1.0000	
6	-0.4018	0.2770	0.0384	-0.0659	0.1590	1.0000

HAZARD OF CHRONIC GVHD   N=163

****
* ESTIMATED MODEL *   GRID INTERVAL :   1.000000
********************   -2 * LOG (LIKELIHOOD) :   576.6478

PARAMETER NUMBER	COVAR. INDEX	COVARIATE NAME	PARAMETER ESTIMATE	EST STD DEV	EXP(PARAMETER)	PARAM/ST.ERR.
FIXED COVARIATES :						
1	0	CONSTANT	-8.3768	0.53917	2.3015E-04	-15.536
2	18	donor age (>20)	0.96244	0.39360	0.3819	-2.4452
3	19	female/male	0.72230	0.33287	2.0592	2.1699
PRESET TIME DEPENDENT COVARIATES :						
4	1	1(T_A<t)	0.60674	0.34326	1.8344	1.7676
5	3	1(T_C<t)	0.50374	0.37145	1.6549	1.3561
COMPUTED TIME DEPENDENT COVARIATES :						
6	1	Z20 (t) = 1(0<=t<150)	1.3886	0.4522	4.0094	3.0704
7	2	Z21 (t) = 1(150<=t<250)	1.3032	0.5212	3.6812	2.5002

*** ESTIMATED ASYMPTOTIC CORRELATION MATRIX OF PARAMETERS ***

	1	2	3	4	5	6	7
1	1.0000						
2	-0.5303	1.0000					
3	-0.2451	-0.0016	1.0000				
4	-0.2426	0.1450	-0.0208	1.0000			
5	-0.5081	0.3933	-0.0524	-0.0768	1.0000		
6	-0.7486	0.2173	0.0023	-0.0422	0.2414	1.0000	
7	-0.5706	0.1129	-0.0024	0.0058	0.0572	0.6336	1.0000

HAZARD OF RELAPSE   N=163

```
*****************
* ESTIMATED MODEL *                GRID INTERVAL :           1.000000
*****************                  -2 * LOG (LIKELIHOOD) :   494.0057
```

PARAMETER NUMBER	COVAR. INDEX	COVARIATE NAME	PARAMETER ESTIMATE	EST STD DEV	EXP(PARAMETER)	PARAM/ST.ERR.
FIXED COVARIATES :						
1	0	CONSTANT	-7.9606	0.50656	3.4892E-04	-15.7150
2	16	under relapse	0.6088	0.51530	1.8383	1.1815
3	17	after 1. remiss.	2.1370	0.51669	8.4740	4.1359
4	18	donor age <20	-1.8295	0.44443	0.1604	-4.1166
PRESET TIME DEPENDENT COVARIATES :						
5	1	1(T_A<t)	-0.5071	0.44504	0.6021	-1.1397
6	2	1(T_G*CMV+<t)	-2.5045	0.81054	8.1716E-02	-3.0899
7	3	1(T_C<t)	-1.7403	0.58363	0.1754	-2.9819
COMPUTED TIME DEPENDENT COVARIATES :						
8	1	Z18 (t) = 1(250<t<=600)	0.83916	0.47726	2.3144	1.7583
9	2	Z19 (t) = 1(t>600)	0.96269	0.55306	2.6187	1.7407

*** ESTIMATED ASYMPTOTIC CORRELATION MATRIX OF PARAMETERS ***

	1	2	3	4	5	6	7	8	9
1	1.0000								
2	-0.0149	1.0000							
3	-0.7788	-0.1645	1.0000						
4	-0.1187	-0.0200	-0.2023	1.0000					
5	-0.0844	-0.1902	-0.1190	0.1103	1.0000				
6	-0.0172	0.0952	-0.1374	0.3567	0.0158	1.0000			
7	-0.1021	0.1083	-0.1158	0.3420	0.0463	0.1572	1.0000		
8	-0.3053	0.0295	0.2115	-0.4277	-0.0865	-0.3054	-0.2435	1.0000	
9	-0.1952	0.0157	0.1314	-0.4451	-0.0457	-0.3246	-0.2826	0.4751	1.0000

HAZARD OF DEATH IN REMISSION   N=163

****
* ESTIMATED MODEL *   GRID INTERVAL :   1.000000
**********************   -2 * LOG (LIKELIHOOD) :   623.7603

PARAMETER NUMBER	COVAR. INDEX	COVARIATE NAME	PARAMETER ESTIMATE	EST STD DEV	EXP(PARAMETER)	PARAM/ST.ERR.
FIXED COVARIATES :						
1	0	CONSTANT	-6.59010	0.69140	1.3738E-03	-9.5316
2	16	under relapse	0.94717	0.43881	2.5784	2.1585
3	17	after 1. remission	0.74296	0.36321	2.1021	2.0455
4	18	donor age (>20)	0.31609	0.36605	0.7290	-0.8635
PRESET TIME DEPENDENT COVARIATES :						
5	1	$1(T\_A<t)$	2.2130	0.36802	9.1427	6.0131
6	2	$1(T\_G<t)$	5.4787E-02	0.42327	1.0563	0.1294
7	3	$1(T\_C<t)$	1.2491	0.37333	3.4870	3.3457
COMPUTED TIME DEPENDENT COVARIATES :						
8	1	$Z02(t) = LOG(t)$	-0.58948	0.12930	0.5546	-4.5590

*** ESTIMATED ASYMPTOTIC CORRELATION MATRIX OF PARAMETERS ***

	1	2	3	4	5	6	7	8
1	1.0000							
2	-0.1820	1.0000						
3	-0.3879	-0.2706	1.0000					
4	-0.1307	0.0911	-0.1361	1.0000				
5	-0.3022	-0.0642	-0.0299	0.1072	1.0000			
6	0.1428	0.0754	0.0137	0.1151	0.0448	1.0000		
7	-0.1703	0.1403	0.1369	0.3660	0.0186	-0.0649	1.0000	
8	-0.7057	0.1336	0.0758	-0.1958	-0.1313	-0.3523	-0.2751	1.0000

HAZARD OF SEPARATION: N=3176   (55-HELSINKI COHORT)

****
* ESTIMATED MODEL *   GRID INTERVAL :   30.00000
*********************   -2 * LOG (LIKELIHOOD) :   3252.161

			PARAMETER			
PARAMETER NUMBER	COVAR. INDEX	COVARIATE NAME	ESTIMATE	EST STD DEV	EXP(PARAMETER)	PARAM/ST.ERR.

FIXED COVARIATES :

1	0	CONSTANT	-2.7618	0.37510	6.317E-02	-7.3628
2	4	Social group	0.9955	0.16994	2.7062	5.8581
3	5	Marital status	2.1581	0.14881	8.6545	14.502

COMPUTED TIME DEPENDENT COVARIATES :

| 4 | 1 | Z02(t) = LOG(t) | -1.0449 | 3.739E-02 | 0.35173 | -27.946 |

*** ESTIMATED ASYMPTOTIC CORRELATION MATRIX OF PARAMETERS ***

1	1.0000			
2	-0.7814	1.0000		
3	0.0385	-0.2472	1.0000	
4	-0.5894	0.0065	0.0567	1.0000
	1	2	3	4

HAZARD OF CHILD PSYCHIATRIC INCIDENCE : N=3176

```
*******************
* ESTIMATED MODEL *        GRID INTERVAL :           30.00000
*******************        -2 * LOG (LIKELIHOOD) :   6308.774
```

PARAMETER NUMBER	COVAR. INDEX	COVARIATE NAME	PARAMETER ESTIMATE	EST STD DEV	EXP(PARAMETER)	PARAM/ST.ERR.
FIXED COVARIATES :						
1	0	CONSTANT	-12.9600	0.51994	2.353E-06	-24.9250
2	4	Social group 55	0.4960	0.10381	1.6422	4.7783
3	5	Change in family 69	0.8128	0.13426	2.2542	6.0539
PRESET TIME DEPENDENT COVARIATES :						
4	1	$1(T\_S<t)$	0.6777	0.17021	1.9695	3.9821
COMPUTED TIME DEPENDENT COVARIATES :						
5	1	Z28 (T) = 1(AGE(7-11))	5.8287	0.50378	339.9100	11.5700
6	2	Z29 (T) = 1(AGE(12-14))	4.6595	0.53427	105.5800	8.7211
7	3	Z42 (T) = 1(AGE(0-6))*T)	2.1448E-03	0.8121E-04	1.0021	7.6270
8	4	Z43 (T) = 1(AGE(7-11)*T)	-3.1012E-04	0.1500E-03	0.9996	-2.6965
9	5	Z44 (T) = 1(AGE(12-15)*T)	-1.2393E-05	0.5053E-06	0.9999	-4.9067E-02
10	6	Z53 (T) = 1(t>T_S)*T_S/30	1.0578E-02	0.3919E-02	1.0107	3.1481

*** ESTIMATED ASYMPTOTIC CORRELATION MATRIX OF PARAMETERS ***

	1	2	3	4	5	6	7	8	9	10
1	1.0000									
2	-0.3231	1.0000								
3	-0.0433	0.0281	1.0000							
4	0.0213	-0.1377	-0.1062	1.0000						
5	-0.9213	-0.0009	-0.0052	-0.0099	1.0000					
6	-0.8701	0.0032	-0.0001	-0.0062	0.8973	1.0000				
7	-0.8884	-0.0010	-0.0058	-0.0089	0.9177	0.8652	1.0000			
8	-0.0036	0.0099	0.0145	0.0094	-0.1860	0.0000	-0.0004	1.0000		
9	0.0011	-0.0031	0.0052	0.0229	-0.0004	-0.3379	-0.0006	0.0011	1.0000	
10	-0.0069	0.0294	-0.0811	-0.5740	0.0085	0.0008	0.0129	-0.0227	-0.0505	1.0000

HAZARD OF PSYCHIATRIC ADMISSIONS: N=3176

```
***********************
* ESTIMATED MODEL *            GRID INTERVAL :              30.00000
***********************        -2 * LOG (LIKELIHOOD) :      3411.652
```

PARAMETER NUMBER	COVAR. INDEX	COVARIATE NAME	PARAMETER ESTIMATE	EST STD DEV	EXP(PARAMETER)	PARAM/ST.ERR.
		FIXED COVARIATES :				
1	0	CONSTANT	-10.747	0.4706	2.1501E-05	-22.8370
2	4	Social group	6.7820E-02	0.1528	1.0702	0.4436
3	6	Change in family 69	0.6339	0.2056	1.8851	3.0835
		PRESET TIME DEPENDENT COVARIATES :				
4	1	$1(t>T\_S)$	0.1278	0.2974	1.1363	0.4297
5	4	$1(t>T\_C)$	1.4433	0.6707	4.2346	2.1517
		COMPUTED TIME DEPENDENT COVARIATES :				
6	1	$Z50(T) = 1(AGE(6-15))$	1.4239	0.4402	4.1534	3.2348
7	2	$Z51(T) = 1(AGE(15-21))$	2.3693	0.4272	10.6890	5.5454
8	3	$Z52(T) = 1(AGE(21-28))$	2.1397	0.4314	8.4972	4.9601
9	4	$Z53(T) = 1(t>T\_S)*T\_S/30$	1.1228E-02	3.1470E-03	1.0113	3.5678
10	5	$Z54(T) = 1(t>T\_C)*EXP(-0.05(t-T\_C/365)$	-1.4911	1.0928	0.2251	-1.3645

*** ESTIMATED ASYMPTOTIC CORRELATION MATRIX OF PARAMETERS ***

	1	2	3	4	5	6	7	8	9	10
1	1.0000									
2	-0.4926	1.0000								
3	-0.0579	-0.0003	1.0000							
4	0.0285	-0.1180	-0.1197	1.0000						
5	0.0262	-0.0372	-0.0633	-0.0243	1.0000					
6	-0.8073	0.0063	0.0040	-0.0082	0.0536	1.0000				
7	-0.8334	0.0084	0.0104	-0.0005	-0.0010	0.8934	1.0000			
8	-0.8268	0.0097	0.0163	-0.0017	-0.1180	0.8776	0.9147	1.0000		
9	0.0314	-0.0189	0.0083	-0.6584	-0.0428	-0.0026	-0.0222	-0.0158	1.0000	
10	-0.0092	0.0089	0.0266	-0.0017	-0.9589	-0.0692	-0.0254	0.0913	0.0075	1.0000

Index

Aalen, O., 25, 28.
Absolute vs. relative time scale, 84, 94-95.
Andersen, P.K., 50, 51, 56, 57, 58.
Arjas, E., 2, 32, 34, 35, 36, 43, 47, 55, 99, 103.
Armitage, P., 57.
Automatic search procedures, 23.
Barlow, W., 99.
Blalock, H., 20.
Bayes' theorem, 10.
Bone marrow transplantation data, 60-79.
    causal chain, 61.
    data, 63.
    hazard models, 66-67.
    cumulative probabilities, 68-70.
    prediction probabilities, 71-74.
    innovation gains, 75-79.
    dependence on history, 74.
Bowlby, J., 80.
Brémaud, P., 32.
Cain, K., 99.
Casella, G., 52.
Causal chain, 1, 29, 31, 34, 43, 55, 61, 80.
Causal effect, 40, 107.
    average/population level, 14, 28.
    causative, 31, 56.
    constant, 59, 66, 71, 74.
    cumulative, 37, 58, 94.
    non-monotonic, 59.
    occurrence time dependent, 95.
    preventive, 31.
    protective, 102.
    transient, 43, 59, 94.
Causality, connotations of, 1.
    deterministic, 4.
    explanatory, 1.
    historical aspects of, 2-3.
    means-end aspect of, 1.
    predictive, 1, 29.
    probabilistic, 4-5.
    qualitative theory of, 6-7.
    quantitative theory of, 5-6.
    regularity theory of, 3.
    statistical analysis of, 13.
Causal field, 3, 109.
Causal lag, 27, 59.
Causal modelling, 2, 7, 19.
Causal process, 11, 19, 24, 27, 109.
Causal relationship, 3, 8, 12, 15, 18, 26, 108.
    unit level, 27, 110.

    unit-treatment effect, 27, 110.
Causal risk difference, 16, 28.
Causal transmission, 29, 30, 42.
Cause, direct/indirect, 7, 20, 22.
    prima facie, 7.
    spurious/genuine, 7, 20, 22.
Chamberlain, G., 24.
Chiang, C., 57.
Change (in prediction), 26, 27, 110.
Conditioning history, 24, 25, 32, 33, 35, 36, 37, 68, 89.
Confidence statements, 44-56.
    confidence bands, 48-51.
    confidence limits, 44, 47-48, 51, 74, 89.
    delta-method, 44, 47-48, 56.
    likelihood-based, 56.
    Scheffé-method, 52.
Counterfactuals, 3, 5, 12-13, 14, 17, 27, 41.
Counting process, 25, 32, 35, 45, 49, 50, 58, 106, 107, 111.
    compensator of, 25, 32, 33, 36, 53, 54.
    stochastic intensity of, 25, 26.
Cox, D.R., 13, 19, 22, 27.
Cox model, 51, 58-59.
Davis, W., 11.
Dempster, A., 13, 24.
Dependence,
    conditional independence, 19, 22, 24.
    dependence chain, 21.
    local, 25.
    marginal independence, 19, 22.
    modelling of, 58-59.
Diggle, P., 108.
Doob-Meyer decomposition, 25.
Ducasse, H., 3.
Dynamic models, 24-25.
    local dependence, 25.
Eells, E., 30.
Eerola, M., 2, 34, 36, 80, 83, 103, 104.
Efron, B., 50.
Elaboration, 20.
Endogenous/exogenous variables, 19.
Eskola, A., 1.
Event, generic vs. singular, 2.
Event chains, see causal chain.
Event-histories, 2, 26.

Exposure, 15, 17, 37.
Filtration, 25, 32.
deFinetti, B., 10, 12.
Fisher, R., 14.
Fleming, T., 49.
Florens, J., 24.
Galai, N., 106.
Giere, R., 8.
Gill, R., 48, 49, 51.
Good, I.J., 5, 6, 11, 15, 27, 30, 34.
Granger, C., 12, 24, 25, 26, 28.
Granger-causality, 24.
Graphical models, 18, 20-23.
Greenland, S., 15, 31, 104.
Hacking, I., 8.
Hall, P., 48.
Hazard model, 37, 45, 58-59, 63, 66-67, 85-86.
    additive, 37.
    multiplicative, 63, 66-67, 85-86.
Helsinki-cohort data,
    causal chain, 80-81.
    data, 81-83.
    hazard models, 83-86.
    survival probabilities, 89.
    prediction probabilities, 89-91.
    innovation gains, 92.
    dependence on history, 91-93.
Hoem, J., 60.
Holland, P., 11, 13, 14, 15, 16, 19, 26, 110.
Hume, D., 2, 3, 5, 6, 28, 56, 109.
Humphreys, P., 9.
Independence models, 18-23.
    graphical models, 18, 20-23.
    chain graphs/dependence chains, 21.
    recursive models, 19, 20, 21.
Informative censoring, 107-108.
Innovation gain, 36-37, 39-43, 48, 70, 66-74, 89.
    confidence limits of, 48, 51, 74.
Interaction, 20, 21, 22, 23, 31, 67, 91, 101, 103.
Jacobsen, N., 61, 63.
Joint distribution, 21, 22, 23.
Kalbfleisch, J., 60.
Kant, I., 2, 3.
Kaplan-Meier estimator, 49.
Keiding, N., 56, 57, 58.
Kiiveri, H., 19.

Klein, J., 57.
Kreiner, S., 23.
Kyburg, H., 7.
Laplace, P., 4.
Lauritzen, S., 21, 22, 23.
Learning effect, 34.
Logistic regression model, 44-46.
    confidence limits in, 51-52.
    hazard of, 45, 66-67, 85-86.
    prediction probabilities in, 46.
    survival function in, 67.
Log-linear model, 21-22.
Longini, I., 57.
Lönnqvist, B., 61.
Mackie, J., 3.
Manton, K., 106.
Marked point process, 31-37, 106.
    canonical space of, 32.
    cause-specificic hazard, 33.
    (crude) hazard of, 33, 106-107.
    history process, 32, 53.
    (sub)distribution of, 32.
Markov assumption, 25, 28, 30, 51, 58.
Martingale, 25, 32, 36, 49, 101.
    residuals, 100-102.
    transform, 49, 50.
Matthews, D., 56.
Mellor, D., 4.
Miettinen, O., 30, 31.
Mill, J.S., 3.
von Mises, R., 8.
Mixed association models, 22-23.
Moolgavkar, S., 57.
Multistate model, 57.
    multiple decrement model, 57.
    multihit model, 57.
    competing risk model, 57, 66-68.
Mykland, P., 25.
Nagel, E., 4.
Nair, V., 48.
Nelson-Aalen estimator, 49, 50, 61.
Norros, I., 32, 34, 35, 36, 43, 53, 54, 55.
Partial regression coefficient, 19, 21.
Partial covariances, 22.
Path analysis, 19, 20.
Pepe, M.S., 57, 58.
Piegorsch, W., 53.
Popper, K., 8.

Positive statistical relevance, 30.
Prediction process, 33-37, 42-43.
   jumps of, 35-36.
   martingale representation of, 36.
Prediction probabilities, 2, 34, 38, 45-46, 67, 86, 96.
   asymptotic variance of, 48, 89.
Prediction method, 35-37, 60, 80, 101, 103, 108, 109, 111.
   event of prediction, 34, 37, 61, 67, 80.
   prediction interval, 37, 41-42, 52, 71, 91, 98.
   time of prediction, 62, 89.
   updating of prediction, 34-36.
Prentice, R., 57.
Principle of indifference, 9.
Probability,
   epistemic, 7, 9-11.
      coherence, 10.
      exchangeability, 10.
      subjective, 9-11, 110.
   physical/objective, 7, 8-9, 11, 12, 109.
   in causality, 7-8, 11-12, 26, 110.
Propensity, 8-9.
Quantitative causal factors, 105-107.
   critical levels of, 106.
   diffusion processes, 107.
   mean level changes in, 106-107.
Randomization, 6, 107.
   confounding, 13, 17-19, 27, 28, 109.
Ramsey, F., 10.
Reichenbach, H., 8.
Relative frequencies, 8, 12.
Robins, J., 13, 15, 16, 18, 19, 31, 104.
Rosenbaum, P., 16.
Rothman, K., 30.
Rubin, D., 11, 13, 14, 16, 19, 27, 28, 60, 110
Rubin's model, 12, 14-16.
Salmon, W., 9, 11, 27, 30.
Savage, L., 10.
Scheffé, H., 52.
Schoenfeld, D., 99.
Schweder, T., 25.
Sensitivity analysis, 93-104.
   of innovation gains, 95-98.
   of regression coefficients, 94-95.
   of transformations, 98-100.
Shaked, M., 53.
Skyrms, B., 11, 12.
Spirtes, P., 23.

Stalnaker, R., 12.
Stochastic covariates, 45-46.
   in causal chain models, 45-46, 66-67, 85-86.
Stochastic order, 53.
   compensator representation, 53.
   of failure times, 53-56.
   monotonic hazards, 51, 54-56.
   supportivity, 54, 56.
Stochastic process, 2, 6, 24, 25, 31-37.
   adapted, 32.
   distribution-valued, 34.
   predictable, 32, 49, 50.
   realizations of, 24, 57.
Structural equation model, 19.
Study designs, 16, 60, 110.
   ignorable, 16, 60.
   noninformative observational plan, 60.
   observational studies, 13, 16, 18.
   randomized experiments, 4, 13, 14-19, 60, 110.
Suppes, P., 4, 5, 6, 9, 11, 12 15, 20, 24, 25, 27, 28, 30, 34, 35.
Systems weakened by failures, 43.
Tennant, C., 80.
Therneau, T., 99.
Transition probabilities, 34, 37.
   see prediction probabilities
Tuma, N.B., 26.
Weiden, P., 57.
Venn, J., 8.
Wermuth, N., 21, 22, 23.
Whittaker, J., 19.
Wiener, N., 24.
Working, H., 52.
Wright, J., 52.
Väisänen, E., 80.
Yarrow, L., 80.
Yule, G., 19.
Zwaan, F., 60.

Lecture Notes in Statistics

For information about Volumes 1 to 8
please contact Springer-Verlag

Vol. 9: B. Jørgensen, Statistical Properties of the Generalized Inverse Gaussian Distribution. VI, 188 pages, 1981.

Vol. 10: A.A. McIntosh, Fitting Linear Models: An Application of Conjugate Gradient Algorithms. VI, 200 pages, 1982.

Vol. 11: D.F. Nicholls and B.G. Quinn, Random Coefficient Autoregressive Models: An Introduction. V, 154 pages, 1982.

Vol. 12: M. Jacobsen, Statistical Analysis of Counting Processes. VII, 226 pages, 1982.

Vol. 13: J. Pfanzagl (with the assistance of W. Wefelmeyer), Contributions to a General Asymptotic Statistical Theory. VII, 315 pages, 1982.

Vol. 14: GLIM 82: Proceedings of the International Conference on Generalised Linear Models. Edited by R. Gilchrist. V, 188 pages, 1982.

Vol. 15: K.R.W. Brewer and M. Hanif, Sampling with Unequal Probabilities. IX, 164 pages, 1983.

Vol. 16: Specifying Statistical Models: From Parametric to Non-Parametric, Using Bayesian or Non-Bayesian Approaches. Edited by J.P. Florens, M. Mouchart, J.P. Raoult, L. Simar, and A.F.M. Smith, XI, 204 pages, 1983.

Vol. 17: I.V. Basawa and D.J. Scott, Asymptotic Optimal Inference for Non-Ergodic Models. IX, 170 pages, 1983.

Vol. 18: W. Britton, Conjugate Duality and the Exponential Fourier Spectrum. V, 226 pages, 1983.

Vol. 19: L. Fernholz, von Mises Calculus For Statistical Functionals. VIII, 124 pages, 1983.

Vol. 20: Mathematical Learning Models — Theory and Algorithms: Proceedings of a Conference. Edited by U. Herkenrath, D. Kalin, W. Vogel. XIV, 226 pages, 1983.

Vol. 21: H. Tong, Threshold Models in Non-linear Time Series Analysis. X, 323 pages, 1983.

Vol. 22: S. Johansen, Functional Relations, Random Coefficients and Nonlinear Regression with Application to Kinetic Data, VIII, 126 pages, 1984.

Vol. 23: D.G. Saphire, Estimation of Victimization Prevalence Using Data from the National Crime Survey. V, 165 pages, 1984.

Vol. 24: T.S. Rao, M.M. Gabr, An Introduction to Bispectral Analysis and Bilinear Time Series Models. VIII, 280 pages, 1984.

Vol. 25: Time Series Analysis of Irregularly Observed Data. Proceedings, 1983. Edited by E. Parzen. VII, 363 pages, 1984.

Vol. 26: Robust and Nonlinear Time Series Analysis. Proceedings, 1983. Edited by J. Franke, W. Härdle and D. Martin. IX, 286 pages, 1984.

Vol. 27: A. Janssen, H. Milbrodt, H. Strasser, Infinitely Divisible Statistical Experiments. VI, 163 pages, 1985.

Vol. 28: S. Amari, Differential-Geometrical Methods in Statistics. V, 290 pages, 1985.

Vol. 29: Statistics in Ornithology. Edited by B.J.T. Morgan and P.M. North. XXV, 418 pages, 1985.

Vol 30: J. Grandell, Stochastic Models of Air Pollutant Concentration. V, 110 pages, 1985.

Vol. 31: J. Pfanzagl, Asymptotic Expansions for General Statistical Models. VII, 505 pages, 1985.

Vol. 32: Generalized Linear Models. Proceedings, 1985. Edited by R. Gilchrist, B. Francis and J. Whittaker. VI, 178 pages, 1985.

Vol. 33: M. Csörgo, S. Csörgo, L. Horváth, An Asymptotic Theory for Empirical Reliability and Concentration Processes. V, 171 pages, 1986.

Vol. 34: D.E. Critchlow, Metric Methods for Analyzing Partially Ranked Data. X, 216 pages, 1985.

Vol. 35: Linear Statistical Inference. Proceedings, 1984. Edited by T. Calinski and W. Klonecki. VI, 318 pages, 1985.

Vol. 36: B. Matérn, Spatial Variation. Second Edition. 151 pages, 1986.

Vol. 37: Advances in Order Restricted Statistical Inference. Proceedings, 1985. Edited by R. Dykstra, T. Robertson and F.T. Wright. VIII, 295 pages, 1986.

Vol. 38: Survey Research Designs: Towards a Better Understanding of Their Costs and Benefits. Edited by R.W. Pearson and R.F. Boruch. V, 129 pages, 1986.

Vol. 39: J.D. Malley, Optimal Unbiased Estimation of Variance Components. IX, 146 pages, 1986.

Vol. 40: H.R. Lerche, Boundary Crossing of Brownian Motion. V, 142 pages, 1986.

Vol. 41: F. Baccelli, P. Brémaud, Palm Probabilities and Stationary Queues. VII, 106 pages, 1987.

Vol. 42: S. Kullback, J.C. Keegel, J.H. Kullback, Topics in Statistical Information Theory. IX, 158 pages, 1987.

Vol. 43: B.C. Arnold, Majorization and the Lorenz Order: A Brief Introduction. VI, 122 pages, 1987.

Vol. 44: D.L. McLeish, Christopher G. Small, The Theory and Applications of Statistical Inference Functions. VI, 124 pages, 1987.

Vol. 45: J.K. Ghosh (Ed.), Statistical Information and Likelihood. 384 pages, 1988.

Vol. 46: H.-G. Müller, Nonparametric Regression Analysis of Longitudinal Data. VI, 199 pages, 1988.

Vol. 47: A.J. Getson, F.C. Hsuan, {2}-Inverses and Their Statistical Application. VIII, 110 pages, 1988.

Vol. 48: G.L. Bretthorst, Bayesian Spectrum Analysis and Parameter Estimation. XII, 209 pages, 1988.

Vol. 49: S.L. Lauritzen, Extremal Families and Systems of Sufficient Statistics. XV, 268 pages, 1988.

Vol. 50: O.E. Barndorff-Nielsen, Parametric Statistical Models and Likelihood. VII, 276 pages, 1988.

Vol. 51: J. Hüsler, R.-D. Reiss (Eds.), Extreme Value Theory. Proceedings, 1987. X, 279 pages, 1989.

Vol. 52: P.K. Goel, T. Ramalingam, The Matching Methodology: Some Statistical Properties. VIII, 152 pages, 1989.

Vol. 53: B.C. Arnold, N. Balakrishnan, Relations, Bounds and Approximations for Order Statistics. IX, 173 pages, 1989.

Vol. 54: K.R. Shah, B.K. Sinha, Theory of Optimal Designs. VIII, 171 pages, 1989.

Vol. 55: L. McDonald, B. Manly, J. Lockwood, J. Logan (Eds.), Estimation and Analysis of Insect Populations. Proceedings, 1988. XIV, 492 pages, 1989.

Vol. 56: J.K. Lindsey, The Analysis of Categorical Data Using GLIM. V, 168 pages, 1989.

Vol. 57: A. Decarli, B.J. Francis, R. Gilchrist, G.U.H. Seeber (Eds.), Statistical Modelling. Proceedings, 1989. IX, 343 pages, 1989.

Vol. 58: O.E. Barndorff-Nielsen, P. Blæsild, P.S. Eriksen, Decomposition and Invariance of Measures, and Statistical Transformation Models. V, 147 pages, 1989.

Vol. 59: S. Gupta, R. Mukerjee, A Calculus for Factorial Arrangements. VI, 126 pages, 1989.

Vol. 60: L. Györfi, W. Härdle, P. Sarda, Ph. Vieu, Nonparametric Curve Estimation from Time Series. VIII, 153 pages, 1989.

Vol. 61: J. Breckling, The Analysis of Directional Time Series: Applications to Wind Speed and Direction. VIII, 238 pages, 1989.

Vol. 62: J.C. Akkerboom, Testing Problems with Linear or Angular Inequality Constraints. XII, 291 pages, 1990.

Vol. 63: J. Pfanzagl, Estimation in Semiparametric Models: Some Recent Developments. III, 112 pages, 1990.

Vol. 64: S. Gabler, Minimax Solutions in Sampling from Finite Populations. V, 132 pages, 1990.

Vol. 65: A. Janssen, D.M. Mason, Non-Standard Rank Tests. VI, 252 pages, 1990.

Vol. 66: T. Wright, Exact Confidence Bounds when Sampling from Small Finite Universes. XVI, 431 pages, 1991.

Vol. 67: M.A. Tanner, Tools for Statistical Inference: Observed Data and Data Augmentation Methods. VI, 110 pages, 1991.

Vol. 68: M. Taniguchi, Higher Order Asymptotic Theory for Time Series Analysis. VIII, 160 pages, 1991.

Vol. 69: N.J.D. Nagelkerke, Maximum Likelihood Estimation of Functional Relationships. V, 110 pages, 1992.

Vol. 70: K. Iida, Studies on the Optimal Search Plan. VIII, 130 pages, 1992.

Vol. 71: E.M.R.A. Engel, A Road to Randomness in Physical Systems. IX, 155 pages, 1992.

Vol. 72: J.K. Lindsey, The Analysis of Stochastic Processes using GLIM. VI, 294 pages, 1992.

Vol. 73: B.C. Arnold, E. Castillo, J.-M. Sarabia, Conditionally Specified Distributions. XIII, 151 pages, 1992.

Vol. 74: P. Barone, A. Frigessi, M. Piccioni, Stochastic Models, Statistical Methods, and Algorithms in Image Analysis. VI, 258 pages, 1992.

Vol. 75: P.K. Goel, N.S. Iyengar (Eds.), Bayesian Analysis in Statistics and Econometrics. XI, 410 pages, 1992.

Vol. 76: L. Bondesson, Generalized Gamma Convolutions and Related Classes of Distributions and Densities. VIII, 173 pages, 1992.

Vol. 77: E. Mammen, When Does Bootstrap Work? Asymptotic Results and Simulations. VI, 196 pages, 1992.

Vol. 78: L. Fahrmeir, B. Francis, R. Gilchrist, G. Tutz (Eds.), Advances in GLIM and Statistical Modelling: Proceedings of the GLIM92 Conference and the 7th International Workshop on Statistical Modelling, Munich, 13-17 July 1992. IX, 225 pages, 1992.

Vol. 79: N. Schmitz, Optimal Sequentially Planned Decision Procedures. XII, 209 pages, 1992.

Vol. 80: M. Fligner, J. Verducci (Eds.), Probability Models and Statistical Analyses for Ranking Data. XXII, 306 pages, 1992.

Vol. 81: P. Spirtes, C. Glymour, R. Scheines, Causation, Prediction, and Search. XXIII, 526 pages, 1993.

Vol. 82: A. Korostelev and A. Tsybakov, Minimax Theory of Image Reconstruction. XII, 268 pages, 1993.

Vol. 83: C. Gatsonis, J. Hodges, R. Kass, N. Singpurwalla (Eds.), Case Studies in Bayesian Statistics. XII, 437 pages, 1993.

Vol. 84: S. Yamada, Pivotal Measures in Statistical Experiments and Sufficiency. VII, 129 pages, 1994.

Vol. 85: P. Doukhan, Mixing: Properties and Examples. XI, 142 pages, 1994.

Vol. 86: W. Vach, Logistic Regression with Missing Values in the Covariates. XI, 139 pages, 1994.

Vol. 87: J. Møller, Lectures on Random Voronoi Tessellations. VII, 134 pages, 1994.

Vol. 88: J.E. Kolassa, Series Approximation Methods in Statistics. VIII, 150 pages, 1994.

Vol. 89: P. Cheeseman, R.W. Oldford (Eds.), Selecting Models From Data: Artificial Intelligence and Statistics IV. X, 487 pages, 1994.

Vol. 90: A. Csenki, Dependability for Systems with a Partitioned State Space: Markov and Semi-Markov Theory and Computational Implementation. X, 241 pages, 1994.

Vol. 91: J.D. Malley, Statistical Applications of Jordan Algebras. VIII, 101 pages, 1994.

Vol. 92: M. Eerola, Probabilistic Causality in Longitudinal Studies. VII, 133 pages, 1994.